CHALLENGES FACED BY TECHNICAL AND SCIENTIFIC SUPPORT ORGANIZATIONS (TSOs) IN ENHANCING NUCLEAR SAFETY AND SECURITY

The Agency's Statute was approved on 23 October 1956 by the Conference on the Statute of the IAEA held at United Nations Headquarters, New York; it entered into force on 29 July 1957. The Headquarters of the Agency are situated in Vienna. Its principal objective is "to accelerate and enlarge the contribution of atomic energy to peace, health and prosperity throughout the world".

PROCEEDINGS SERIES

CHALLENGES FACED BY TECHNICAL AND SCIENTIFIC SUPPORT ORGANIZATIONS (TSOs) IN ENHANCING NUCLEAR SAFETY AND SECURITY

STRENGTHENING COOPERATION AND IMPROVING CAPABILITIES

PROCEEDINGS OF AN INTERNATIONAL CONFERENCE ORGANIZED BY THE
INTERNATIONAL ATOMIC ENERGY AGENCY,
HOSTED BY THE GOVERNMENT OF CHINA
THROUGH THE NATIONAL NUCLEAR SAFETY ADMINISTRATION,
THE STATE NUCLEAR SECURITY TECHNOLOGY CENTER,
CHINA ATOMIC ENERGY AUTHORITY,
THE RADIATION MONITORING TECHNICAL CENTER OF
THE MINISTRY OF ENVIRONMENTAL PROTECTION,
SUZHOU NUCLEAR SAFETY CENTER,
SUZHOU NUCLEAR POWER RESEARCH INSTITUTE,
BEIJING REVIEW CENTER OF NUCLEAR SAFETY
IN COOPERATION WITH THE
EUROPEAN TECHNICAL SAFETY ORGANISATIONS NETWORK (ETSON)
AND HELD IN BEIJING, CHINA, 27–31 OCTOBER 2014

INTERNATIONAL ATOMIC ENERGY AGENCY
VIENNA, 2018

COPYRIGHT NOTICE

© IAEA, 2018

Printed by the IAEA in Austria
November 2018
STI/PUB/1833

IAEA Library Cataloguing in Publication Data

Names: International Atomic Energy Agency.
Title: Challenges faced by Technical and Scientific Support Organizations (TSOs) in enhancing nuclear safety and security / International Atomic Energy Agency.
Description: Vienna : International Atomic Energy Agency, 2018. | Series: Proceedings series (International Atomic Energy Agency), ISSN 0074–1884 | Includes bibliographical references.
Identifiers: IAEAL 18-01196 | ISBN 978–92–0–108118–6 (paperback : alk. paper)
Subjects: LCSH: Nuclear industry — Safety measures. | Nuclear industry — Security measures. | International cooperation.
Classification: UDC 621.039.58 | STI/PUB/1833.

FOREWORD

The Global Nuclear Safety and Security Framework (GNSSF) provides a conceptual structure and guidelines for achieving and maintaining a high level of safety and security at nuclear facilities and in nuclear related activities around the world. Technical and scientific support organizations (TSOs) play an essential role in sustaining the GNSSF by providing assistance to regulatory bodies in establishing and maintaining nuclear and radiological programmes with a strong safety and security component built on a sound technical and scientific basis.

Since 2007, the IAEA has held a series of international conferences examining the role played by TSOs in their support of regulatory bodies and the nuclear industry. The first conference, held in Aix-en-Provence, France, in 2007, provided a forum for TSOs, international organizations and experts to discuss — and develop a common understanding of — the roles, responsibilities, needs and opportunities of TSOs. The second conference in the series, held in 2010 in Tokyo, Japan, focused on international cooperation and networking among TSOs, and the roles of TSOs in the regulatory process and in capacity building in Member States considering embarking on a nuclear power programme.

The third conference in this series, the International Conference on Challenges Faced by Technical and Scientific Support Organizations (TSOs) in Enhancing Nuclear Safety and Security: Strengthening Cooperation and Improving Capabilities, held in Beijing, China, in October 2014, was convened in the wake of the accident at the Fukushima Daiichi nuclear power plant. Thus a primary objective was to examine TSOs and their role in the light of the Fukushima Daiichi accident. Through the presentations and discussions, the conference participants also assessed the effectiveness of TSOs and explored ways to improve capabilities and strengthen cooperation among TSOs. Other topics addressed included the challenges faced by TSOs when interacting with regulatory bodies, the industry and the public; the roles of TSOs in terms of emergency preparedness and response; maintaining and strengthening TSO capabilities; and networking among TSOs in a global environment.

Over 200 participants from 42 Member States and 5 organizations attended the conference. During the meeting 3 keynote presentations and 33 invited papers were given. Each of the first six sessions was accompanied by a panel discussion which allowed additional speakers to give statements.

This publication includes summaries of the conference sessions, the opening addresses, abstracts of the invited papers, and the Conference President's report, including the conclusions and deliberations of the meeting. The attached CD-ROM contains the invited papers and presentations, the contributed papers and their respective posters, and the list of participants.

The IAEA wishes to thank the contributors involved in the preparation of this publication. The IAEA officers responsible for this publication were P. Woodhouse, L. Guo, M. Heitsch and K. Ben Ouaghrem of the Office of Safety and Security Coordination.

EDITORIAL NOTE

CONTENTS

SUMMARY 1

OPENING SESSION 2

THE ROLE OF TSOs IN RELATION TO THE FUKUSHIMA DAIICHI
ACCIDENT (TOPICAL SESSION 1)... 27

INTERFACE ISSUES (TOPICAL SESSION 2) 37

EMERGENCY PREPAREDNESS AND RESPONSE (TOPICAL SESSION 3).................. 43

MAINTAINING AND STRENGTHENING TSO CAPABILITIES
(TOPICAL SESSION 4)... 51

NETWORKING AMONG TSOs AND BEYOND (TOPICAL SESSION 5) 59

CLOSING SESSION 67

SUMMARY AND CONCLUSIONS OF THE CONFERENCE 69

ACKNOWLEDGEMENTS 78

CONFERENCE ORGANIZATION 79

ANNEX: CONTENTS OF THE ATTACHED CD-ROM 81

SUMMARY

The International Conference on Challenges Faced by Technical and Scientific Support Organizations (TSOs) in Enhancing Nuclear Safety and Security: Strengthening Cooperation and Improving Capabilities was held in Beijing, China, in October 2014, built on two previous conferences on this subject, held in Tokyo, Japan, in 2010, and in Aix-en-Provence, France, in 2007. The focus of this third TSO conference was on strengthening cooperation among TSOs and improving their capabilities to provide nuclear and radiation safety and security expertise to regulatory bodies and operators. The overall aim was to assess and review ways to further improve the effectiveness of TSOs, taking into account lessons learned from the accident at the Fukushima Daiichi nuclear power plant in March 2011.

The Fukushima Daiichi accident had a direct impact on the daily work of TSOs, and the role of TSOs increased with the implementation of the IAEA Action Plan on Nuclear Safety (the Action Plan). Adopted by the IAEA's Board of Governors and endorsed by the IAEA General Conference in September 2011, the Action Plan set out a comprehensive programme of work, in 12 major areas, to strengthen nuclear safety worldwide in response to the accident. This third TSO conference was organized to share the lessons to be learned from the accident and to assess the impact of the accident on the work of TSOs. The conference also provided a forum for discussion of the roles, functions and importance of TSOs in enhancing nuclear and radiation safety and nuclear security, including through capacity building in those countries launching or expanding their nuclear power programme.

A goal of the conference was to facilitate the exchange of experience and good practices in planning and implementing cooperative activities for capacity building and in identifying needs for assistance activities from the standpoint of recipient countries. The role of TSOs in the implementation of the Action Plan was also highlighted. Specific objectives were to:

— Consider appropriate approaches to enhancing cooperation and effective networking among TSOs and beyond, including the creation of centres of excellence;

— Provide an overview of the technical and scientific support needed to maintain a sustainable nuclear safety and security system;

— Foster dialogue on all relevant technical, scientific, organizational and legal aspects at the international level.

Benoît de Boeck, Director General of Bel V, Belgium, served as the Conference President. A total of 209 participants from 42 countries and five organizations, as well as two observers, participated in the conference; ten press representatives also attended.

The conference programme comprised an opening session, five topical sessions that focused on challenges faced by TSOs in enhancing nuclear safety and security, and a concluding panel session that focused on future developments and future cooperation among TSOs.

For each session a short summary is given followed by abstracts of the invited papers.

OPENING SESSION

The opening session began with a welcome address by Li Ganjie, Vice Minister of Environmental Protection and Administrator of the National Nuclear Safety Administration of China, who introduced the role of the TSO in China. This was followed by two opening addresses: D. Flory, Deputy Director General of the Department of Nuclear Safety and Security of the IAEA, reiterated the value of collaboration on the challenges facing TSOs and described the role of TSOs in capacity building and the support provided to TSOs by the IAEA. B. De Boeck discussed demands on TSOs and regulatory bodies worldwide, as well as the impact of the Fukushima Daiichi accident on TSOs and TSO networks, and on the role played by TSOs. The opening session also included three keynote presentations: J. Repussard, Director General of the Institute for Radiological Protection and Nuclear Safety (IRSN), France, reported on progress in the implementation of recommendations from the last TSO conference, held in 2010. G. Caruso, Special Coordinator for the IAEA Nuclear Safety Action Team, reported on progress in the implementation of the Action Plan and the IAEA report on the Fukushima Daiichi accident. M. Hirano, Director-General for Regulatory Standard and Research at the Nuclear Regulation Authority of Japan, summarized the activities related to the Fukushima Daiichi accident being carried out in Japan.

Note: The titles and abstracts of all keynote addresses have been included for completeness of the publication.

OPENING ADDRESS

Li Ganjie

Vice Minister,
Ministry of Environmental Protection,
and Administrator of NNSA,
Beijing, China

Distinguished Deputy Director General Mr. Flory, Distinguished Chairman B. De Boeck, ladies and gentlemen,

Good morning!

The International Conference of technical and scientific support organizations (TSOs) is an important platform for enhancing the capacity of TSOs in nuclear safety and promoting international exchange and cooperation in this regard. Two sessions have been organized before and in-depth discussions were held on topics including enhancing the technical and scientific support capacity of TSOs and enhancing the international cooperation, with a series of outcomes achieved. Since the 1st TSO conference, we have developed a more common understanding on the status, function, importance and basic principles of TSOs. All countries have further developed their TSOs, and technical support and resource are further enhanced. A regular liaison mechanism has been established among TSOs, enabling effective joint research, experience sharing and peer reviews. All these achievements have contributed greatly to strengthening nuclear safety regulation in different countries and raising global nuclear safety to a new level.

The 3rd TSO conference is a major event in the international nuclear safety sector, and it is also a big occasion for nuclear safety regulation in China. This conference will, building on the previous two conferences, draw lessons from the Fukushima Accident, increase our common understandings, further identify responsibilities, seize the opportunities and play our roles, and improve the efficacy of TSOs. The conference follows the trend of international nuclear development and falls in line with the theme of nuclear safety regulation, thus it is of important historical relevance.

On the occasion of the conference, on behalf of the Ministry of Environmental Protection/National Nuclear Safety Administration (MEP/NNSA), I'd like to extend my warmest welcome to the participating delegates and guests from all countries, and express my sincere gratitude to people from all walks of life that support and care for nuclear safety. My thanks also go to diligent colleagues working at TSOs.

Ladies and gentlemen,

The thriving of TSOs has become a strong boost for the continuous improvement of global nuclear safety. The nuclear energy development and nuclear technology application in China has moved into fast track, and the role of TSOs are highly recognized.

1. The TSOs in China have experienced fast growth and played major role in ensuring nuclear safety.

China has set up an all-inclusive organizational structure for TSOs, with clear hierarchy in place. The NNSA has set up five permanent TSOs, namely, the Nuclear and Radiation Safety Center (NSC), the Radiation Monitoring Technical Center, the Suzhou Nuclear Safety Center, China Productivity Center for Machinery, and Beijing Nuclear Safety Review Center. Radiation TSOs are also available in all the provincial environmental protection bureaus and some of the prefectural-level environmental protection agencies.

The professional team engaged in nuclear safety regulation in China comprises of 100 staff at NNSA, 1,000 at the organizations of central level, and 10,000 at local levels. After three decades of review and regulatory practice of nuclear safety, this team has accumulated abundant engineering and administrative experience, with an echelon of professionals at different age groups, who are well-educated and specialized in all relevant fields.

The Chinese Government has always taken it seriously to providing financial support for TSOs. In recent years, resource input in nuclear safety regulation has been growing. Since 2011, project funding allocated to TSOs in China has been growing by 28% every year, and reached 210 million RMB in 2014, accounting for 74% of the total funding. Even in the context of reduced government expenditure in recent years, the funding for TSOs of nuclear safety still has been on the increase year by year.

China is now constructing the National Research & Development Base for Nuclear and Radiation Safety Regulation, in an effort to reinforce the technical support capacities in nuclear safety review, supervision, monitoring, emergency response, publicity campaign, and international cooperation. The projected acreage of the Base is 218 mu (14.53 ha.). The Phase-I building area is 93,000 m2 with 750 million RMB investment. The total building area is close to 200,000 m2. Upon its completion, the Base will be developed into a leading international platform for integrated and specialized technical support for nuclear safety regulation.

The TSOs in China are capable of offering technical consultation and service in terms of nuclear safety policies, plans, laws and regulations, standards; nuclear safety review, inspection, environmental impact assessment; emergency response and environmental monitoring; qualification management of professionals; and public outreach.

The TSOs in China have offered scientific and technical support for the government's regulatory decision making and policy making, playing an essential role in ensuring national nuclear safety. In terms of legal framework, the TSOs have studied China's legal system on nuclear safety, developed a five-year legislation plan, followed the international nuclear safety standard, and provided expert suggestions; in terms of nuclear safety regulatory regime, the TSOs have carried out in-depth researches on the challenges in nuclear safety regulation under the current conditions, and come up with the implementation plans for achieving modernization of nuclear safety regulation in China; In terms of emergency preparedness and response, the TSOs have undertaken the responsibilities of sustaining and maintaining nuclear and radiation

4

emergency response capacities and offering technical support for national emergency response; in terms of international exchange and cooperation, the TSOs have sent out experts to take part in the technical exchange and cooperation in bilateral and multilateral programs. Furthermore, the TSOs have offered technical support for the evaluation of Pakistan nuclear safety programs. In terms of nuclear safety planning, the TSOs have identified primary challenges faced by nuclear safety regulation in the medium and long term, prepared a draft nuclear safety planning, and helped with the assessment of implementation effect; in terms of public outreach, the TSOs have studied the information distribution mechanism, promoted the information disclosure to the public, and timely responded to public opinions; in terms of technical research and development, the TSOs have carried out overarching design, facilitated the technical breakthroughs in key areas, and promoted the application of scientific research findings. With regard to the nuclear safety examination and improvements in the aftermath of Fukushima Accident, the TSOs have supported the safety evaluation of nuclear facilities and studied the safety requirements for new nuclear power plants and generic technical requirements for the safety improvements in nuclear power plants.

The nuclear safety regulation in China has been greatly enhanced thanks to the effective operation and strong support of TSOs. The authority and effectiveness of nuclear safety regulation is secured.

2. The TSOs in China have gained valuable experience that shall be summarized and upheld

First, we shall stick to the fundamental role of science and technology.

Nuclear safety is complicated in technological terms. It involves all fields of technology and management in this industrialized society. It is imperative that all fields shall be effectively organized in the collaborations and cooperation. Nuclear accidents are usually emergency cases, which may occur and progress quickly within a very short period of time. Professional judgment is required for rapid response; the consequence of nuclear accidents is difficult to cope with. Specialized professionals are needed to carry out the accident investigation, emergency rescues and environmental restoration; nuclear safety is a socially sensitive matter. Specialized professionals are needed for publicity campaign and authoritative interpretation. Specialized TSOs can be depended upon to effectively mobilize professionals in conducting nuclear safety regulation. This is the common practice and useful experience of all countries in nuclear safety regulation. It is also an effective model for governments to introduce modernized management.

Second, we shall stick to the principle of independence of technical support services.

Independence is a fundamental principle of nuclear safety regulation. Maintaining the independence of technical support is the inevitable requirement for this principle in nuclear safety regulation. As the largest TSO in China, the Nuclear and Radiation Safety Center (NSC) is affiliated to MEP/NNSA and undertakes nuclear safety review; NNSA has always paid attention to any possible interest connections between a TSO and the licensees. NNSA is prudent in allocating review program, in an effort to prevent possible interest-related interferences with the review conclusion.

3. We shall introduce competitive mechanism for technical support services.

Introducing competitive mechanism helps to ensure the quality of technical services, enable sound resource allocation, and contribute to human resources nurturing and technical advance. On one hand, we are determined to expand and strengthen the TSOs as the cornerstones; on the other, we shall introduce competitive mechanism to give full play to the supplementary role of other TSOs through long-term and effective cooperation. In practice, we resort to market-based bidding process to choose the best and most competitive bidders for drafting regulations and standards, conducting R&D, developing policy study and engaging review program. We also implement an A/B Party review system, in which two TSOs will be selected to carry out back-to-back review regarding key issues, in an effort to encourage competition and ensure review quality.

4. We shall stick on matching technical support services with international practice.

Nuclear safety knows no boundary. Technical exchange shall have no barriers. The TSOs in China have paid full attention to studying and learning from the latest international standards on nuclear safety, and improving China's nuclear safety regulatory system and legislations and standards. They have also conducted extensive information exchange, technical discussions and joint researches with international counterparts to share state-of-art concepts, methodology, technologies and experience; the TSOs in China have received regular peer reviews and made improvement of the technical support system; use international resources for training through the "bring in" and "go global" strategies.

The nuclear power is now undergoing a new round of rapid growth, and the public has increasingly high demand for safety. The TSOs will undertake heavier tasks, shouldered with more challenging work and play an important role. Hereby I would like to propose that, the TSOs of all countries should 1) improve its verification, calculation, testing and validation capacity, and build up nuclear safety evaluation and review center; 2) make greater efforts in the R&D of nuclear safety technologies, and build up nuclear safety technology R&D and application center; 3) improve the information technology of nuclear safety, and build up data collection and exchange center; and 4) establish a sound human resource nurturing mechanism, and foster human resources incubator.

Ladies and gentlemen,

This great conference is held at a time when NNSA is celebrating its 30th anniversary. Three decades of meticulous efforts have enabled us to build up a solid shield in nuclear safety, and three decades of unremitting explorations have blazed out a trail of glorious history of nuclear safety. The 30 years of NNSA can be divided into three stages.

The first stage is the "startup exploration" from 1984 to 1998, beginning with the official establishment of National Nuclear Safety Administration on July 2, 1984. Under the management of State Science and Technology Commission, the Shanghai, Guangdong, Chengdu, and North China Nuclear Safety Inspection Offices were gradually set up, followed by other TSOs like Beijing Nuclear Safety Center, Suzhou

Nuclear Safety Center, Beijing Nuclear Safety Review Center under Beijing Institute of Nuclear Engineering, and Technical Research Center of Nuclear Equipment Safety and Reliability under the China Academy of Machinery Science and Technology. We began the nuclear safety regulation with lawmaking by promulgating the Regulations on Supervising the Safety of Civilian Nuclear Facility. We drafted laws and standards in line with international practice and applicable to Chinese conditions. We also abide by the principle of governance by law, and introduced and implemented some basic regulatory systems, such as nuclear safety licensing system, and supervision and inspection system.

The second stage is "integration and improvement" from 1998 to 2008 beginning with the incorporation of NNSA into State Environmental Protection Administration (SEPA) in March 1998. China promulgated and enforced the Law on Prevention and Control of Radioactive Pollution in 2003, which is the first of its kind in the field of nuclear safety. The regulatory organizational structure has been further improved. Six regional offices were set up, covering all regions in China. NSC, the Radiation Monitoring Technical Center and other TSOs have been either expanded or upgraded with enhanced technical forces. Moreover, a permanent and stable team was completed with several technical support bodies, including China Institute for Radiation Protection, China Institute of Atomic Energy, Tsinghua University, and a dozen of other scientific research institutions and higher educational institutions.

The third stage is the "rapid growth" since 2008, beginning with the upgrading of SEPA to Ministry of Environmental Protection, which also acts as the National Nuclear Safety Administration. In 2011, NNSA is expanded into three administrative departments from only one department before. The institutional of nuclear safety regulation has made a leap-forward, with optimized functions, growing workforces, improved legal system, and enhanced regulatory capacity.

Nuclear safety regulation in China has made great achievements over the past 30 years. We started from scratch, we make innovation and work unswervingly to keep up with time and reap in greater achievement.

1) Infrastructure development. NNSA has developed one culture, one system, one team and a group of capabilities. "One culture" refers to the nuclear safety culture of "defense in depth, continuous improvement, innovation and high efficiency, fairness and sequence, strictness, prudence, carefulness and practicality". "One system" refers to the regulations and institution system. "One team" refers to the regulatory organization covering NNSA headquarters, regional offices and TSOs. "A group of capabilities" refer to the capacity in areas such as license review, supervision and law enforcement, radiation monitoring, emergency response, experience feedback, technical R&D, public outreach and international cooperation.

2) License review. NNSA has implemented license management on civil nuclear facilities including 20 in-service nuclear power units, 28 under-construction units, 19 research reactors and criticality devices and 20 nuclear fuel cycle facilities. In addition, it also ensures license management of radiation safety on more than 110,000 in-use radioactive sources and over 120,000 ray-emitting devices.

3) Inspection and law enforcement. NNSA conducts 24-hour inspection on various kinds of nuclear facilities across the country. All important safety-related activities are included in its routine inspection. Life-cycle supervision on radioactive sources is

implemented. NNSA never tolerates and severely punished any conduct in violation of nuclear safety regulations, with great results achieved.

4) Monitoring and emergency response. NNSA has set up a national radiation monitoring network that releases monitoring data in time. It has improved national system for emergency response to nuclear and radiation accident, developed emergency preparedness plans and carried out drills on regular basis.

With these efforts, the safety performance of all in-service nuclear power units across the country is sound. The frequency of radiation accident of radioactive sources has gone down from 6.2 accidents per 10,000 sources every year in 1990s to less than 1 accident per 10,000 sources every year.

With common understanding, firm belief, unremitting exploration, and orderly progression, China has accumulated precious experience in nuclear safety regulation over the past 30 years.

First, we follow the basic rule on nuclear safety regulation. We always adhere to the basic policy of "safety first, quality foremost" to cultivate systematic safety culture, set up quality assurance system and implement defence in depth concept. Second, we not only follow international practice, but also cater to Chinese conditions of China. We state that nuclear safety is an integral part of national security; considering nuclear safety as the biggest economic benefits and as the lifeline for the development of nuclear energy and application of nuclear technologies. Third, we adhere to the basic principle of supervision according to law. We adhere to strict law enforcement to ensure the authority and effectiveness of nuclear safety regulation. Fourth, we attach importance to capacity building in nuclear safety regulation. We have continuously improved our capacity in all areas such as review, inspection, monitoring emergency response, experience feedback, research and development, public outreach and international cooperation.

Ladies and gentlemen,

We had a magnificent and brilliant history of nuclear safety regulation over the past 30 years. At this new starting point, we are faced with new situation, new tasks and new requirements. We should keep on with our effort, make innovations, consolidate our foundation and strengthen support. We should work hard on modernization of the regulatory system and capacity; we should work to improve regulations and organizational structure, enhance technical capacity; enrich cultural development and make contributions to sound, safe and sustainable development of nuclear energy and nuclear technologies.

Innovation is the driving force for achieving modernization of nuclear safety regulation. The first is theoretical innovation. A theoretical system for regulation that is highly applicable to Chinese conditions will be fostered, with culture as the background, management as the guide and technology as the bases: the second is institutional innovation. We will adhere to the principle of independence to integrate regulatory functions and set up a relatively complete regulatory institution for nuclear safety and build up a nuclear safety regulatory system suitable to the development of nuclear energy and technologies in China. The third is mechanism innovation. We will improve the mechanisms such as technical review, supervision and law enforcement, monitoring, emergency response, research and development, management of regulations and public outreach and we will improve supervision efficiency. The fourth is technical innovation. We will accelerate the development of technical innovation system, increase the inputs in scientific research on nuclear safety, concentrate on key study projects and promote scientific progress of the whole industry.

The traditional buildings in China usually include four beams and eight pillars. Modern architecture also stresses on the mechanic combination of foundation and supporting structures. The nuclear regulatory system is like a mansion that requires solid foundation to ensure its firm stand and indestructible structure. Similarly, the nuclear safety regulation is also a "mansion" that has four cornerstones and eight supporting structures, or four beams and eight pillars.

Legal framework, institution and workforce, technical capacity as well as culture are the four cornerstones of the mansion of nuclear safety regulation in China.

Facing the future, firstly, we should consolidate the cornerstone of legal framework. We will draft Nuclear Safety Law as soon as possible and improve the top-down design of laws in nuclear field, including Atomic Energy Law, Nuclear Safety Law and Law on the Prevention and Control of Radioactive Pollution. Secondly, we will consolidate the cornerstone of institution and workforce. We will facilitate the establishment of NNSA with complete institutional arrangement (still affiliated to MEP). Thirdly, we should consolidate the cornerstone of technical capacity. We will develop National Technology Research and Development Base for Nuclear and Radiation Safety Regulation and complete the three platforms of independent analysis, testing and verification, information sharing as well as exchanges and training. Fourthly, we should consolidate the cultural cornerstone. We will implement the nuclear security concept proposed by President Xi Jinping on nuclear safety featuring "rationality, coordination and synchronized progress"; we will popularize nuclear safety culture; and enhance the awareness to the big picture, risk, progress and rules.

The Chinese government has always attached great importance to nuclear safety and nuclear safety culture. China has always engaged in nuclear energy development and nuclear technologies, with safety as the prerequisite. In order to implement China's nuclear security concept and national security strategy, promote the overall improvement of national security and ensure the sound, safe, and sustainable development of nuclear energy and technologies, NNSA is going to release the policy statement on nuclear safety culture based the experience and best practice of China in the development of nuclear safety culture for the past 30 years. The statement will expound our position on nuclear safety culture, describe eight features of nuclear safety culture, present proposal for creating sound cultural atmosphere in the entire industry and call for the whole society to improve nuclear safety. This is a major move of China in nuclear safety regulation and has great significance and relevance.

There are eight supporting structures (pillars) as below for nuclear safety regulation in China, which need to be further enhanced in the future.

1) Review and license. We will vigorously facilitate the reform of administrative license according to the State requirement to streamline the administration and decentralization; we will set up a review methodology suitable for nuclear safety regulation in China and improve our review capacity.

2) Inspection and law enforcement. We will strengthen inspection during and after any procedure. We will enhance the capacity building in inspection and law

enforcement. We will optimize the inspection and examination methodology and improve the technical level of inspection and examination.

3) Radiation monitoring. We will improve the national radiation monitoring of environment quality, supervisory monitoring system for nuclear facilities and emergency response monitoring system for radiation environment. We will improve the distribution of radiation monitoring network and optimize the layout of monitoring sites under national or provincial monitoring program.

4) Emergency response. We will develop a professional team for emergency response under unified command and management. We will set up an expert review system, carry out emergency response drills; enhance local capacity in emergency response and improve emergency response in local governments.

5) Experience feedback. We will improve filing system, create information platform, set up expert pool for experience feedback and improve NNSA's experience feedback system.

6) Research and development. We will promote the construction of National Technical Research and Development Base for Nuclear and Radiation Safety Regulation, improve the management system on scientific research, cultivate research team, ensure continuous input and enlarge the applications of relevant research results.

7) Public outreach. We will focus on standardizing and institutionalizing public outreach on nuclear safety. We will work out a system of information disclosure on nuclear facilities, enhance public participation and improve the public relation and response system for nuclear emergencies.

8) International cooperation. We will strengthen international cooperation and learn from advanced experience. We will closely follow the development trend of nuclear safety around the world, draw from the advanced experience in management and inspection from other countries and promote continuous improvement of nuclear safety regulation in China.

Ladies and gentlemen,

The International Atomic Energy Agency (IAEA) is the advocator and organizer of international cooperation on nuclear safety. It provides a great platform for international cooperation in this regard. It has facilitated continuous improvement of nuclear safety around the world and offered strong support for China's nuclear safety regulation. The cooperation between China and IAEA will be further expanded and the international exchanges between China and other countries will become deeper and wider. China will support IAEA as always, we will continue to strengthen international cooperation on nuclear safety, we will promote information exchanges and communications, encourage experience sharing and feedback, implement emergency response and assistance, share technologies and experience and make greater contribution to enhancing global nuclear safety.

This Conference provides a great platform for professionals of all countries to share their successful experience. I believe that, through in-depth exchanges and discussions this week, all participants will have a deeper understanding of the status, function, and the important roles of TSOs. It is expected that joint studies, experience sharing and peer reviews will be effectively carried out at wider range. I also hope that through this Conference, all participants would understand more about China's nuclear safety and China itself. and provide more support for China's development. Let us work hand in hand to create a strong nuclear safety defense line through science and technology and jointly promote the progress of global nuclear safety through trust and cooperation.

Finally, I wish a great success of the 3rd TSOs Conference. I wish you a pleasant stay here and good health!

Thank you!

OPENING ADDRESS

D. Flory

Deputy Director General,
Department of Nuclear Safety and Security
International Atomic Energy Agency,
Vienna

Good morning, Mr. President and friend Benoit de Boeck, Minister and friend Li Gan Jie, Ladies and Gentlemen, distinguished guests and colleagues.

On behalf of the International Atomic Energy Agency, it gives me great pleasure to welcome you all to this *2014 International Conference on Challenges Faced by Technical and scientific Support Organizations (TSOs) in Enhancing Nuclear Safety and Security*. I would like to extend my sincere appreciation to the Government of China and in particular, to the Ministry of Environmental Protection and the National Nuclear Safety Administration (NNSA), the Chinese regulatory body, for hosting this important conference in the historically fascinating city of Beijing — a city which is also experiencing tremendous development, growth and change.

The meeting was organized in close cooperation with ETSON, the European TSO network. A highlight of the conference will be the celebration of the 30th anniversary of NNSA.

During the next days we will have more than 30 invited presentations in the five topical sessions defined by the Programme Committee. Due to time restrictions the more than 40 contributed papers can only be presented in poster sessions during the breaks. Please visit the poster sessions and discuss major statements and findings with the authors.

When I opened the previous TSO conference in Tokyo, 4 years ago in October 2010, I had been in office just one and a half month, and nobody knew that less than 5 months later, the international nuclear community – operators, regulators, vendors, but also Governments and international organizations, would go through one of their greatest challenges, the accident at the Fukushima Daiichi Nuclear Power Plant. In my introductory speech, then, I said that "the introduction of new nuclear power plants in newcomer countries, the rapid expansion of existing nuclear power programmes and the wider use of radioactive sources and ionizing radiation, in general highlight the need for continued and improved international cooperation, strengthened capacity building and infrastructure development and knowledge networking to address the associated challenges." This is still valid today, though the word **rapid,** qualifying the expansion of nuclear programmes can apply only to some countries, such as our host today.

Let me contribute to setting the stage for this week by first describing how I see the character and value of this important conference in which we have all invested a commitment of our precious time, knowledge and experience. I will try to open the discussion, on some of the difficult challenges TSOs face, and I will finally highlight how the Agency has and will continue to support and promote TSOs roles and their continued use in all countries—developed and developing alike.

Value of collaboration

First, these periodic conferences provide a forum for discussion and broad interaction on the future direction for international collaboration amongst TSOs with the strong promotional support of the IAEA. This week, we have the opportunity to add deeper insight and clearer foresight to the role of the TSO in nuclear safety and security worldwide through our active participation, information sharing and strong questioning. Our return on investment in this conference will be measured by the outcomes we achieve this week that can indeed be implemented to strengthening the role of TSOs and their global coordination and collaboration, especially in countries in the process of expanding or embarking on a nuclear programme.

Difficult challenges and opportunities for progress

Since 11 March 2011, the impact that the Fukushima nuclear accident has had on nuclear safety and on the use of nuclear power in general has led to renewed scrutiny of plant design and siting, defence in depth, severe accident management, radiation protection, waste safety regulations as well as emergency response measures. It has also impacted upon some countries' policies on energy and their decisions about the development of nuclear power.

In the wake of the Fukushima accident, IAEA Member States unanimously endorsed the IAEA Action Plan on Nuclear Safety. One very important outcome of the Action Plan, is the extension of the role of the Agency to provide Member States, international organizations and the general public with timely, clear, factually correct, objective and easily understandable information during a nuclear emergency on its potential consequences, including **analysis of available information and prognosis of possible scenarios** based on evidence, scientific knowledge and the capabilities of Member States.

Such a role cannot be understood without a closer cooperation between the IAEA and technical and scientific organizations able to conduct such analysis and prognosis. We have already started strengthening both our own capacities and our cooperation in this field with TSOs.

While the accident at Fukushima did not stop the expansion of nuclear power, it did, once again, highlight the challenges associated with safety and security underlying its development. However, increasing energy demands, coupled with concerns over climate change and the use of fossil fuels continue to be drivers in the expansion of nuclear power and we continue to see more new reactors under construction.

Training experts and harmonizing technical assessment approaches for nuclear facilities based on the exchange of best practices constitute challenges as well as paths for progress in the field of nuclear safety, radiation protection and nuclear security.

Conventions and amendments

For a safe nuclear development programme, building and maintaining national and international trust is a necessity. Part of this trust relies on being respectful of the international legally binding instruments developed along the years, surprisingly usually following major events.

The accident at the Chernobyl NPP was at the origin of the Convention on Nuclear Safety, the CNS. Today, following the Fukushima Daiichi accident, Contracting Parties are getting prepared for considering a draft amendment proposed by Switzerland. It is certainly too soon to give any prediction on the outcome of the Diplomatic Conference convened on February 9th 2015. However, the subject is tightly linked to the commitment of Contracting Parties expressed at the 2nd Extraordinary Meeting, and in particular that:

"The displacement of people and the land contamination after the Fukushima Daiichi accident calls for all national regulators to identify provisions to prevent and mitigate the potential for severe accidents with off-site consequences. Nuclear power plants should be designed, constructed and operated with the objectives of preventing accidents and, should an accident occur, mitigating its effects and avoiding off-site contamination. The Contracting Parties also noted that regulatory authorities should ensure that these objectives are applied in order to identify and implement appropriate safety improvements at existing plants."

Addressing this subject, either through an amendment to the CNS or through direct commitment of Member States, should certainly be a high priority for designers, operators, and regulators alike. It needs also first and foremost a technical and scientific evaluation of safety improvements designed to address these strengthened objectives.

Capacity building and the role of TSOs

There are three major issues associated with nuclear safety and security today: the first is the availability of Human competences, the second is the availability of Human competences, and I believe you have understood that the third is also the availability of Human competences.

In a nutshell, we are facing a shortage of good safety oriented science, and of bright young specialists committed to safety.

Let us address the shortage of young safety specialists. How do you attract bright young guys to the field of safety expertise? Big money? I cannot speak for everybody, but in the last decade, if you really wanted to make money, you chose business rather than science. I do not think we can compete. So it must be a challenging scientific job. We need to include research in the day to day missions of TSOs. Good news is that if we can go this way, it will at the same time address two of our challenges: science and human resources. At the same time, this will also help strengthen the questioning attitude, a common necessary feature of science and safety.

IAEA promotes TSOs

Nuclear safety, nuclear security, are not administrative issues. In many cases, decisions need to be science based. It is for this reason that within - or aside - Regulatory Bodies, the function of Technical and Scientific Support, has been an essential factor for many years and will become ever more important, in the coming years.

While the primary responsibility for safety lies with the operator, TSOs provide the necessary scientific approaches and support to the authorities, the regulators, and even sometime to the public, in the development of nuclear safety and security. They can provide technical evaluations that underlie regulatory decisions. They can support safety authorities in setting up and enforcing independent and competent regulatory policies. Further areas of support to both regulator and operator include training of staff and knowledge dissemination.

Considering each country's own history and experience in the nuclear field, it is understandable that the role of a TSO is not the same in every country. This is well known and not surprising. What is new, however, is the increasing importance of TSOs across national borders. New countries that do not share the same historical and technological experiences are entering the scene and are looking for support. TSOs have therefore increasingly gained importance in providing assistance to regulatory bodies with limited resources by providing comprehensive competence in all aspects of nuclear safety and security.

As the use of TSOs by the regulator and operators becomes increasingly critical, and their use varies significantly from country to country, both regulators and operators will need to clarify their expectations of the TSOs' functions, as wells as to have in-house competencies to manage, review and verify their TSOs' activities, the 'intelligent customer' concept.

Traditionally many Technical Support Organizations have been focused on support for nuclear safety.

However in many countries this has changed over time, particularly where a single regulatory body has responsibility for safety and security and has access to the services of a Technical Support Organization. At the last international conference on Technical Support Organizations convened in Tokyo in 2010 it was recommended that Technical Support Organization functions be extended to provide technical support to competent authorities in the field of nuclear security.

Already the two first conferences concluded among others that in between the international meetings there should be a forum to exchange information and to organize international collaboration for the international experts.

In 2012, you established under the aegis of the IAEA the TSO Forum, with a focus on exchanging information and improve collaboration between international groups or organizations of experts.

The Steering Committee of the TSO Forum includes currently representatives from 20 countries and from the EC and the OECD/NEA. The IAEA hosts the Forum, and its steering committee will hold its meeting in the margins of the Conference.

The TSO Forum is integrated in the Global Nuclear Safety and Security Network (GNSSN) of the IAEA and benefits from comprehensive information resources. The integration of the Forum enables its members to have access to information portals broader than the scope of the Forum. Since 2012 the TSO Forum has coordinated numerous activities, including:

Drafting a policy document describing the roles, functions and management of TSOs, which will be further developed in cooperation with the Department of Nuclear Energy within the Agency;

Providing a summary of the lessons learned by the expert groups in the individual countries in response to the Fukushima Daiichi accident;

Detailed discussions on "Safety and Security Interfaces in Emergency Situations";

Close cooperation and exchange of working results with the European TSO network ETSON;

Presentation and promotion of the TSO Forum at IAEA General Conferences through dedicated side events.

The IAEA will continue to assist Member States in developing a common understanding of the TSOs' responsibilities, needs and opportunities. We will continue to promote international cooperation and networking between TSOs; and we will continue to foster capacity building through the use of TSOs in countries embarking on nuclear power programmes, and in those in need of building up their human resources.

Before concluding, I would like to take this opportunity to praise the work of the international Programme Committee of the conference, which indeed received much support from the members of the TSO Forum. As you know, the preparation of an international meeting requires many steps and activities, and the scientific secretariat which was initially in the expert hands of Matthias Heitsch, had to be transferred into the evenly expert hands of Lingquan Guo when the evolution of nuclear policies in Germany accelerated Matthias departure to the JRC (he still accepted to lend a hand during this week).

And finally, I want to express my gratitude to the Conference President, Benoit de Boeck, for gracefully accepting, at very short notice, to take over the role of President from Terry Jamieson who unfortunately could not join us this week. As President of the conference he will have the always complex task to summarize the results obtained and may give a perspective for the work of the Technical and Scientific Experts and their organizations in the next years.

Thank you for your attention and I wish you a productive conference and look forward to fruitful discussions on these topics during the week.

OPENING ADDRESS

B. De Boeck

President of the Conference,
General Manager,
BEL V,
Brussels, Belgium

Good morning ladies and gentlemen, dear colleagues, dear friends.

I am very honoured to give the opening address as President of this third International Conference on Challenges Faced by Technical Support Organizations in Enhancing Nuclear Safety and Security. As you may know, the President should have been Terry Jamieson from the Canadian Nuclear Safety Commission. He is unfortunately unable to attend, which he deeply regrets given the time and effort he had dedicated to the preparation of this important event. He was really looking forward to meet you all here. This opening address is in part coming from him and I want to thank him for his input.

In March 2011, less than 5 months after the 2010 TSO conference in Japan, we were again reminded that "an accident anywhere is an accident everywhere". We, as TSOs, were called upon to support our national governments as the events at Fukushima Daiichi unfolded.

The demands on TSOs and regulators world-wide were staggering. During the events, we were required to assess what had happened, what could possibly happen and how this might affect us in our own countries. We were also called upon to determine what needed to be done domestically to verify the safety of our own domestic nuclear installations, and to propose what additional safety improvements were to be recommended. Session 1 is devoted to this subject.

On top of this, many of us were asked to contribute to the communication, and by this I mean communication at all levels, of factual information to address the concerns of our stakeholders and the public. Finally, in the three years since Fukushima, TSOs and regulators have worked to ascertain why this accident happened and to ensure that it could not happen again. The role of the IAEA in this area will be presented a little later this morning.

To perform these tasks, TSOs required technical and scientific knowledge, the ability to perform research and development, access to adequate human and capital resources and the ability to reach science-based conclusions in a timely manner and to communicate them effectively. But these of course are the core attributes of a TSO. The only difference was that in the case of Fukushima, we had to execute our mission 24 hours a day, 7 days a week, and on a real-time basis dictated by the laws of physics. Fortunately, the international TSO community was prepared.

However, this is not to say that there were no lessons to be learned. Some of these lessons were new, whereas others served to refocus our attention.

So, for our discussions over the coming days, we will examine:

— The impact of the Fukushima Daiichi accident on TSOs and TSO networks;

— The interface and communications issue for TSOs;

— The issue of emergency preparedness and response;

— The need to maintain and strengthen the TSO capabilities;

— The added value of the networking of TSOs.

We have made much progress since the first TSO conference in 2007, and in particular since the previous one in 2010, as will be explained by Jacques Repussard in a few moments. We retain our collective goal to further our capacity as Technical Safety Organizations and to leverage our individual capabilities. We understand that this is both necessary and achievable. Indeed, we instinctively worked together during Fukushima, helping each other to interpret data, to arrive at recommendations and collaborating on common messaging.

This conference should also consider how the overall scientific and technical knowledge base, needed to support the regulator, is maintained, increased and shared. This might be done through research, training, operational experience, scientific analysis or technical exchanges. This includes how it can be shared between TSOs and perhaps with special emphasis on support to the TSO for new entrants and those dealing with new or unique technologies in their country.

Our role post-Fukushima is to continue to evaluate nuclear safety, to fully identify and properly assess risks and to help put risks into perspective. This role has not fundamentally changed, but our challenge now is to operationalize our findings in an open and risk-informed manner, and to ensure that high levels of safety and transparency are realized world-wide. I see the role of the TSO Forum as a means to provide global expert support to the IAEA and partner countries in the case of a nuclear accident or radiological emergency.

It is encouraging to see that a proper forum has been established for emergency preparedness and response under the Global Nuclear Safety and Security Network. I would like to acknowledge the GNSSN's contribution for providing a dynamic forum to various international regional and thematic networks for sharing knowledge, good practices and enhancing cooperation.

I was last week at the IAEA in Vienna for a meeting of the International Nuclear Safety Group, the INSAG. Let me share with you two of the issues raised there and for which the contribution of the TSOs may be very helpful. The first one is defence in depth; yes, this old concept, one of the most important foundations of safety, applied also in fields other than nuclear. No need to remind you of the five classical levels; many IAEA documents describe them in detail. But there is an additional and complementary way to implement defence in depth, and this is within the nuclear institutional system, which can be seen as consisting of three independent main barriers:

(i) A strong nuclear industry;

(ii) A strong nuclear regulator;

(iii)A strong set of stakeholders.

Components of the industry barriers are the staff of the licensee, the nuclear safety committee of the utility, the WANO peer reviews and the IAEA OSART missions.

Components of the regulatory barrier are the safety authority, the TSO, the international peer pressure (e.g. the review meetings of the Convention on Nuclear Safety), and the international peer reviews (e.g. the IAEA IRRS missions). Components on the stakeholders' barrier may include the parliament, the local civil authorities, the neighbours, and the media.

I do not have the time to discuss them all here, but I want to emphasize the important role the TSOs have to play in the regulatory barrier. By bringing an independent science based view on nuclear safety, a TSO greatly reinforce the ability of a regulator to play its role in the surveillance of the safety responsibility of the operating organization. It will therefore be necessary to acknowledge and maybe reinforce the role of TSOs in the implementation of institutional defence in depth.

The second issue is the protection against external events. It is well known, and it has been shown during the Fukushima Daiichi accident, that an extreme external event will breach all at once the first four levels of defence in depth, and maybe also weaken the last level: emergency management. Moreover, the frequency of extreme external events is difficult to assess, be it an earthquake, a tsunami, or an external flooding, to name only a few. The paucity of the historical record over the last thousands of years, and the uncertainties in the models needed to predict the magnitude and probability of these events explain why it is so difficult to quantify their frequency. Showing convincingly that this frequency is below 10-4 per year is next to impossible. On the other hand the probability of a large early release caused by an internal event is required to be below 10-6 per year. This may raise the question to know if our way to assess risk is really balanced.

The role of TSOs in the issue of protection against external events starts at site selection by using the best science available to identify potential external hazards and quantify their frequency; and it continues at the design stage by ensuring that the design provisions to resist external events and limit their consequences are sufficient. Finally TSOs have to re-assess this risk at the occasion of the periodic safety reviews to incorporate new scientific knowledge and changes in the initial assumptions of the safety demonstration.

To conclude, and on behalf of the Programme Committee, I would like to extend my warmest thanks to all of you who have contributed to the preparation and now to the conduct of this conference. In particular, I would like to thank the Government of China for making this excellent venue available, and to the IAEA for their administration of the conference. We have many important matters to discuss and debate over the coming days. Please speak up and make your opinions known. It is only by this exchange and dialogue that we can meaningfully advance our collective TSO capability.

Thank you, and I wish you all a productive conference in the marvellous city of Beijing.

KEYNOTE ADDRESS

PROGRESS IN THE IMPLEMENTATION OF RECOMMENDATIONS FROM THE LAST CONFERENCE HELD IN 2010

J. Repussard

Director General of the Institute for Radiological Protection
and Nuclear Safety (IRSN),
France

Abstract

As a follow up to the conclusions of the first Conference held in France, the second TSO Conference in Tokyo in 2010 sought to achieve such objectives as: to develop a common understanding of the responsibilities, needs and opportunities of TSOs; to promote International Cooperation and Networking between TSOs; to foster capacity building and the work of TSOs in countries embarking in nuclear program, or in those with limited as well as extensive experience. To address those issues, the Tokyo Conference recognized the importance for the global safety community to maintain and continuously develop TSO functions, which should be adequately recognized in national regulatory systems. In this context, it was highlighted that TSOs have to rely on the following sources and capabilities: adequate human and financial resources; scientific risk-oriented research; relevant operating experience analysis; capacity building, professional educational and training courses; knowledge management. Those elements still remain keys in ensuring nuclear safety on sound scientific bases. However, six months after the second Conference, the Fukushima accident in March 2011 brought further challenges and generated a global mobilization to enhance safety and radiation protection worldwide. Many new initiatives have emerged, such as the IAEA Action Plan on Nuclear Safety, worldwide stress tests, enhanced legislation and regulation, notably in Europe. Lessons learned also pointed out to the key importance of emergency preparedness and response and of improved communication to the public. Experience showed that all those aspects rely largely on scientific and technical support and highlighted further the critical importance of TSO functions. Moreover, further to the recommendations of the Conference in Tokyo, the 'TSO Forum' was established, with a view to cooperating more effectively and on a regular basis, addressing common challenges and sharing experiences with respect to nuclear safety and security. In addition, synergies between nuclear safety and security were developed, following the principles set out by the IAEA. Four years after the Tokyo Conference, substantial progress has been achieved on many of its recommendations as well as on other challenges. Nevertheless, the implementation of several recommendations still needs efforts, for instance with respect to the development of IAEA documents to define a framework and provide sufficient guidance on the roles and functions of TSOs in ensuring nuclear safety, including its interface with nuclear security, or regarding the implementation of the necessary capability building process in embarking countries to develop required TSO functions, in particular through knowledge and experience transfer by the international community. The paper will elaborate on progress achieved and remaining challenges, illustrating the analysis with concrete examples taken from different countries and from Europe as an entity with its specific regional nuclear safety approach.

KEYNOTE ADDRESS

G. Caruso

Nuclear Safety Action Team,
Department of Nuclear Safety and Security,
International Atomic Energy Agency,
Vienna

Abstract

Approximately 70% of the some 800 tasks in the IAEA Action Plan on Nuclear Safety have been completed. Among the tasks where technical and scientific support organizations (TSOs) provided key support were the safety assessments (stress tests); IAEA peer reviews; the extension of capabilities in RANET; the review of IAEA safety standards; harmonization of the liability regime within the international legal framework. Open and transparent communication is crucial. The IAEA is taking a leading role in putting together the forthcoming Fukushima report. The report will be a factual evaluation and assessment of the accident comprising a summary report, written to be understandable to laypersons, and five scientific/technical chapters. Not all the lessons learned from the accident at the Fukushima Daiichi nuclear power plant were new lessons. When moving forward, steps need to be taken to ensure that past lessons 'stay learned'.

KEYNOTE ADDRESS

SUMMARY OF FUKUSHIMA RELATED ACTIVITIES IN JAPAN

M. Hirano

Director-General for Regulatory Standard and Research,
Nuclear Regulation Authority

Abstract

This keynote presentation presents an overview and update on the new regulatory framework in Japan including merger of the former TSO, JNES (Japan Nuclear Energy Safety Organization) with the regulatory body, the NRA (Nuclear Regulation Authority), and on the Fukushima Daiichi-related activities with a focus on on-site stabilization such as fuel removal from the spent fuel pools and management of large amount of radioactive water toward safe and prompt decommissioning. The NRA was established as an independent and integrated commission body in September 2012 and urgently started developing the new regulatory requirements for nuclear power plants which came into force in July 2013. So far, a total of 20 units, 12 PWRs and 8 BWRs, have applied for conformance review to the new requirements for restart. On March 1, 2014, the former JNES was merged with NRA to enhance the technical competence and expertise of NRA. On that occasion, a new department, Regulatory Standard and Research Department was created in NRA as a so-called "internal TSO" for developing the technical standards and guides and conducting safety research. In parallel, cooperation with the Nuclear Safety Research Center in JAEA (Japan Atomic Energy Agency) and NIRS (National Institute for Radiological Sciences) which are the external TSOs has been strengthened. Regarding Fukushima Daiichi, Tokyo Electric Power Company (TEPCO) has conducting various activities according to the Mid-and-Long-Term Roadmap towards Decommissioning under the supervision of the Council for Decommissioning of TEPCO's Fukushima Daiichi Nuclear Power Station of the government. TEPCO has already started fuel removal from the spent fuel pool (SFP) at unit 4, more than 75% of which has been completed. According to the roadmap, fuel debris removal would start in the first half of the fiscal year 2020 at earliest. It would take 30 to 40 years to complete the whole process. Large amount of radioactive water being created daily is a difficult issue that needs long-term efforts. Highly radioactive water remaining in the seawater pipe trenches in the seaside area is believed to be the highest risk contributor at the moment. In order to drain the radioactive water in the trenches, TEPCO is attempting to plug the flow paths between the trenches and turbine buildings by applying the ice plugging technique that will also be applied for construction of the so-called "frozen soil wall" surrounding the units 1 to 4.

Introduction

This paper presents an overview and update on the new regulatory framework in Japan including merger of the former TSO, the Japan Nuclear Energy Safety Organization (JNES), with the Nuclear Regulation Authority (NRA), as well as those on the Fukushima Daiichi-related activities focusing on its onsite stabilization such as fuel removal from the spent fuel pools and management of large amount of radioactive water toward safe and prompt decommissioning.

Lessons Learned Identified by the Diet's Report

On July 5, 2012, the National Diet's Fukushima Nuclear Accident Independent Investigation Commission (NAIIC) reported its investigation results to the National Diet of Japan (Diet's Report: http://warp.da.ndl.go.jp/info:ndljp/pid/3856371/naiic.go.jp/en/). It is written in the message from the chairman:

"What must be admitted - very painfully - is that this was a disaster "Made in Japan." Its fundamental causes are to be found in the ingrained conventions of Japanese culture: our reflexive obedience; our reluctance to question authority; our devotion to 'sticking with the program'; our groupism; and our insularity...."

Regarding regulatory independence, it is also written:

"actual relationship lacked independence and transparency, ... In fact, it was a typical example of "regulatory capture," ... that means that the regulatory body was actually manipulated by the industries...."

Regarding lack of technical expertise, it is written:

"the two incorporated technical agencies advising NISA (Nuclear and Industrial Safety Agency, the former regulatory body), TSOs, namely, JNES and JAEA (Japan Atomic Energy Agency), have been too rigidly tied to NISA...."

In its conclusions, it is written:

"The lack of expertise resulted in "regulatory capture,"... They avoided their direct responsibilities by letting operators apply regulations on a voluntary basis.

As seen above, an importance of so-called "Technical independence" has been emphasized. Regarding this technical independence, a report (NEA/CNRA/R(2014)3) was issued by the OECD Nuclear Energy Agency Committee on Nuclear Regulatory Activities (CNRA) entitled "The Characteristics of an Effective Nuclear Regulator", which clarified the concept of the technical independence as well as "Political independence" and "Financial independence" as the utmost important elements for being effectively independent from undue influence in decision making as shown below.

Political independence: Authorized and being able to make independent regulatory judgments and regulatory decisions within their field of competence for routine work and in crisis situations.

Financial independence: Provided with sufficient financial resources, reliable funding and staffing for the proper and timely discharge of its assigned responsibilities. ...

Technical independence: Possessing technical and scientific competence and the capacity to make independent decisions; having access to independent scientific and technical support.

Current Status of NRA

In June 2012, the Nuclear Regulation Act (NRA) was amended and the new regulatory framework was set forth. The new framework is characterized by the following:

Measures against large scale natural hazards and criminal acts including terrorism.

New regulations on severe accidents (SAs): Mandatory measures to prevent SAs and mitigate their consequences.

Regulation based on the state-of-the-art knowledge and information: All existing nuclear installations are required to comply with the new regulatory requirements (backfitting).

Legal requirement for 40 year limit of operation for NPPs: The NRA can permit an extension less than 20 years.

Special regulation to disaster-experienced NPPs.

Three months later, in September 2012, the NRA was established as an independent and integrated commission body and the nuclear regulation functions regarding safety, security, radioisotopes and others were integrated into the NRA.

Soon after that, the NRA started developing the new regulatory requirements for nuclear power plants which, in the end, came into force in July 2013. So far, a total of 20 units, 12 PWRs and 8 BWRs, have applied for conformance review to the new requirements.

Since nuclear safety and security are to a great extent scientific in nature, the technical independence is of utmost importance for regulatory decision-making, as already discussed in Sections 2. Then, the former JNES was merged with the NRA on March 1, 2014, and a new department, the Regulatory Standard and Research Department, was created within the NRA as an "internal TSO" for developing the technical standards and guides, and conducting safety research to enhance its technical competence and expertise. In parallel, cooperation with the "external TSOs," the Nuclear Safety Research Center in JAEA and NIRS (National Institute for Radiological Sciences) has been strengthened.

In order to achieve and maintain this technical independence, it is obvious that safety research plays an important role. Regarding the safety research in the NRA, special emphasis has been placed on the external and internal hazards leading to a large scale common cause failure, severe accident phenomena and others as shown below.

— Research on extreme natural hazards:

- Hazard evaluation of earthquake, tsunami and others, fragilities of SSCs, etc.;

- Monitoring of volcanic unrests, etc.;

- Methods and models for probabilistic risk assessments (PRAs) especially those for external and internal fire and floods, multi-hazards, multi-units, application of level 3 PRA, etc.

— Research on SAs:

- Code development for SA progression and source terms, etc.;

- Experimental studies on pool scrubbing under depressurization, effect of seawater injection, spray cooling in loss-of-coolant accidents (LOCAs) at spent fuel pools (SFPs), etc.

— Research on Fukushima Daiichi:

- Management of wastes and contaminated water, risk assessment, etc.;

- Criticality of fuel debris during its retrieval operation, etc.

Other areas: Decommissioning and waste disposal, SAs at fuel cycle facilities, etc.

1. Current Status of Fukushima Daiichi

Regarding Fukushima Daiichi, Tokyo Electric Power Company (TEPCO) has conducting various activities according to the Mid-and-Long-Term Roadmap towards Decommissioning under the supervision of the Council for Decommissioning of TEPCO's Fukushima Daiichi Nuclear Power Station of the government.

TEPCO has already started fuel removal from the spent fuel pool (SFP) at unit 4, more than 75% of which has been completed. According to the roadmap, fuel debris removal would start in the first half of the fiscal year 2020 at earliest. It would take 30 to 40 years to complete the whole process.

A large amount of radioactive water being created daily is a difficult issue that needs long-term efforts. Highly radioactive water remaining in the seawater pipe trenches in the seaside area is believed to be the highest risk contributor at the moment. In order to drain the radioactive water in the trenches, TEPCO is attempting to plug the flow paths between the trenches and turbine buildings by applying the ice plugging technique that will also be applied for construction of the so-called "frozen soil wall" surrounding the units 1 to 4.

2. Summary and Challenges

Based on the lessons learned from the Fukushima Daiichi accident, the NRA was created as an independent and integrated regulatory body in September 2012. Since safety and security are to a great extent scientific in nature, technical Independence is of utmost importance for regulatory decision-making. Then, the former JNES was merged with the NRA on March 1, 2014 to enhance its technical competence and expertise.

Regarding Fukushima Daiichi, a large amount of radioactive water being created daily is a difficult issue that needs long-term efforts. Currently, removal of highly radioactive water remaining in the sea-side trenches is a high priority issue.

TSOs need to timely contribute to resolving regulatory issues with high priority and, at the same time, they need to be vigilant and proactive to new findings and emerging future needs. Therefore, safety research should play a key role. In this context, maintaining technical infrastructure has been emphasized to be important and is still a challenge:

Continuous recruiting and developing skilled research engineers;

Maintaining experimental facilities, hot laboratories, etc.;

Implementing an interface function with natural scientists, for example, in the academia, responding to the glowing needs for natural sciences such as seismology, meteorology and volcanology.

International information exchange and joint research projects organized by such as IAEA, OECD/NEA and ETOSN (European TSO Network) as well as cooperative activities under bilateral arrangements are playing an essential role to achieve a common goal.

THE ROLE OF TSOs IN RELATION TO THE FUKUSHIMA DAIICHI ACCIDENT
(TOPICAL SESSION 1)

Chairperson

M. HIRANO
Japan

The first topical session presented TSO responses to the Fukushima Daiichi accident, including: the challenges faced and the solutions identified; TSO involvement in the implementation of stress tests; assistance provided by TSOs in the formulation and implementation of nuclear safety regulations; TSO responses to the IAEA Action Plan on Nuclear Safety; and the role of TSOs in post-accident recovery. The session included a presentation describing the work of a committee established by Japan's Nuclear Regulation Authority to investigate specific issues. The need for independence on the part of the regulatory body as a key component of safety culture, and the need for regulators to have access to sufficient, technically competent resources, either in-house or through external expert support, was also highlighted. The wide range of activities carried out by the European Technical Safety Organisations Network (ETSON) in response to the Fukushima Daiichi accident were presented; these activities were expected to drive future proposals in R&D and technical safety assessment guidelines. One of the main lessons learned from the first topical session was that many elements of nuclear safety are science based, requiring sufficient technical expertise either within or external to the regulatory body. In this sense, international cooperation remains crucial.

PEER REVIEW AND IMPLEMENTATION PROCESS OF EU STRESS TESTS

A. STRITAR
Ministry of the Environment and Spatial Planning,
Ljubljana, Slovenia

Abstract

The accident at the Fukushima Daiichi Nuclear Power Plant in March 2011 was a milestone in nuclear industry that has once again emphasized the importance of responsible and conservative decision making processes among all stakeholders involved in assuring nuclear safety. The lessons learned are showing how important is the preparedness to worst scenarios by the NPP operators, how crucial is critical review and assessment by the regulatory bodies and certainly how inevitable is excellent understanding of natural phenomena based on the best available knowledge base.

The Fukushima accident in March 2011 represented a big challenge to everybody involved in nuclear safety. The first meeting of everybody involved with nuclear safety in EU was held already 4 days after the tsunami in Japan. The decision was made to start immediate campaign for analysis of vulnerability of European NPPs to external events and for implementation of potential improvements. The initiative was soon endorsed by the European Council. WENRA and ENSREG have prepared comprehensive specifications for so called Stress Tests, which were endorsed at the end of May 2011. During the rest of 2011 operators of all NPPs in EU, Switzerland and Ukraine and national regulators have spent hundreds of man years analysing vulnerabilities of their facilities and preparing improvement measures. By the end of 2011 national reports were made public. In spring of 2012 the intensive Peer Review process was going on where a group of about 80 regulators has reviewed all national reports, visited selected facilities and prepared recommendations to national regulators.

The final report of the Stress Test campaign was endorsed by ENSREG on 25 April 2012 and was later delivered to the European Council. The Peer Review Team has determined that a lot has been done in all NPPs in Europe and that there are also plans for further long term improvements. The report is highlighting four major recommendations on European level (need for development of common reference levels about protection against external hazards, importance of containment integrity, importance of Periodic Safety Reviews and importance of severe accident management preparedness). 17 National Peer Review Reports are introducing additional recommendations to each national regulator. ENSREG has approved a special Action Plan on 1 August 2012 to make sure that the conclusions from the stress tests and their peer review result in improvements in safety across European nuclear power plants. Each country has prepared its own National Action Plan. Their implementation will be cross-checked during another peer-review process planned for the first half of 2013. This will ensure that the recommendations and suggestions from the stress test peer review are addressed by national regulators and ENSREG in a consistent manner.

In parallel with the stress test campaign in 2011 WENRA has started the development of additional Reference Levels addressing issues related to the safety against external events. After almost three years of preparations and discussions with stakeholders new reference levels were approved in July 2014. In next years all European nuclear countries will harmonize their nuclear safety related legislative framework with these Reference Levels.

During the Stress Test campaign, it was becoming more and more obvious that improvements are needed also in the area of emergency preparedness off the nuclear sites, i.e. in the wider surroundings and on national levels. Special task force under the umbrella of WENRA and HERCA is currently working on harmonization nuclear emergency preparedness arrangements in different countries.

DISCUSSION OF SOME NEW SAFETY CONCEPTS AND NEW SAFETY REQUIREMENTS IN LIGHT OF THE FUKUSHIMA ACCIDENT

G. CHAI
Nuclear and Radiation Safety Center,
Beijing, China

Abstract

After Fukushima Nuclear Accident, some new safety concepts and new safety requirements are suggested and discussed among the nuclear industry and nuclear safety regulatory organizations all over the world. In this paper, new safety concepts and new safety requirements, such as "Design Extension Condition", "enhance the application of Defence in Depth", "independence between different levels of Defence in Depth", "enhance the diversity design of safety features", "safety level should be As High As Reasonable Achievable", and "practically elimination of large release of radioactive materials" are discussed; and also it is stated in this paper that, with the consideration of "safety level should be As High As Reasonable Achievable", deterministic and probabilistic methodologies should be used to identify the safety voluntaries in the design of NPPs, and reasonable practicable measures should be taken to minimize the consequence of residual risk, and to achieve the safety goal of practically elimination of large release of radioactive materials.

RESPONSE TO THE IAEA ACTION PLAN ON NUCLEAR SAFETY BY TSO

A. KHAMAZA
Scientific and Engineering Centre for Nuclear and Radiation Safety,
Moscow, Russian Federation

Abstract

Based on the IAEA Action Plan on Nuclear Safety, the Russian Federation developed an "Action Program of Russian Authorities and Organizations Concerned in Implementation of the IAEA Action Plan on Nuclear Safety" up to the year 2015. The paper presents information on the Program implementation by the Russian regulatory authority for nuclear and radiation safety in the field of use of atomic energy (Rostechnadzor), and also considers the results of stress tests conducted for the Russian NPPs (either in operation or planned for construction) and the most powerful research reactors. Moreover, it provides information about R&D and standard-setting activities in the field of regulation of atomic energy use carried out by the Scientific and Engineering Centre for Nuclear and Radiation Safety (SEC NRS – Rostechnadzor's TSO) to support the Rostechnadzor's activity aimed at implementation of the IAEA Action Plan on Nuclear Safety. The paper also emphasizes the results of the IAEA Integrated Regulatory Review Service (IRRS) follow-up mission that was held in November 2013, in preparation and conduct of which SEC NRS took part.

STUDY ON SEVERE ACCIDENT PROGRESSION AND SOURCE TERMS IN FUKUSHIMA DAIICHI NPPs

H. HOSHI,
R. KOJO,
A. HOTTA,
M. HIRANO
Nuclear Regulation Authority,
Tokyo, Japan

Abstract

It has been past three years since the severe accident in TEPCO's Fukushima Daiichi Nuclear Power Plants. After Fukushima Daiichi accident, several investigation reports were published. These reports pointed out lessons learned from the accident. Nuclear Regulation Authority (NRA) enacted new regulations last year, which require accident management and counter measures against severe accident, etc., to enhance safety of nuclear power plants and nuclear facilities. On the other hand, NRA launched new committee to investigate unsolved issues of the accident, which were pointed out previous investigation reports. Decommissioning of severely damaged plants is in progress. In parallel, onsite R&Ds for investigation of the accident are undergoing. However, due to technical difficulties for investigation such as high radiation, leakage of contaminated water, etc., available information, especially for the inside of primary containment vessel, is limited. S/NRA/R is conducting both experimental programs and computational analyses to study important severe accident phenomena and accident progression of Fukushima Daiichi accident. This paper summarizes current research activities.

THE UK AND EUROPEAN RESPONSE TO FUKUSHIMA

A. HALL
Office for Nuclear Regulation,
Bootle, United Kingdom

Abstract

Following the accident at the Fukushima Daiichi site, ONR's Chief Nuclear Inspector reported on the implications and lessons learnt for the UK nuclear industry. This review extended to all UK nuclear licenced sites and concluded that analysis of the Fukushima Daiichi accident revealed no fundamental safety weaknesses in the UK nuclear industry. However, 38 areas were identified where lessons could be learned in the UK from the crisis in Japan. In developing its response to Fukushima, ONR co-operated extensively with international organizations, notably with the IAEA and the Western European Nuclear Regulators Association (WENRA). ONR's approach was consistent with European response to Fukushima, developed by the European Commission and European Nuclear Safety Regulators Group (ENSREG). The report highlighted the importance of the principle of "continuous improvement" to achieving high standards of nuclear safety. This principle is embedded in UK law, where there is a requirement for nuclear designers and operators to reduced risks so far as is reasonably practicable. This is underpinned by the requirement for detailed periodic reviews of safety (throughout the life of an installation) to seek further improvements. This means that, no matter how high the standards of nuclear design and subsequent operation are, the quest for improvement should never stop. Seeking to learn from events, new knowledge and experience, both nationally and internationally, is a fundamental feature of the safety culture in the UK nuclear industry. Whenever a major accident occurs there are, not unreasonably, questions and comments directed to the regulatory body in relation to its role overseeing the safety of facilities. Often questions arise over the independence of the regulator, and the approach taken for individual regulatory decisions. The concept of "intelligent customer" has developed in UK to ensure that organizations account for their legal duties for any work commissioned externally. Applied to the regulatory context, this concept requires regulators to have sufficient competent resource within the regulatory body to specify, oversee and accept support provided under technical support contracts. Regulatory decisions are made by the warranted inspectors to safeguard regulatory independence.

POST FUKUSHIMA RESEARCH IN THE VIEW OF THE EUROPEAN TSO NETWORK ETSON

F.-P. WEISS,
Gesellschaft für Anlagen- und Reaktorsicherheit GmbH,
Cologne, Germany
Email: frank-peter.weiss@grs.de

V. DELEDICQUE
Bel V,
Brussels, Belgium

L. CIZELJ
Jožef Stefan Institute,
Ljubljana, Slovenia

E. SCOTT-DE-MARTINVILLE
Institut de Radioprotection et de Sûreté Nucléaire,
Fontenay-aux-Roses, France

S. RIMKEVICIUS
Lithuanian Energy Insitute,
Kaunas, Lithuania

T. LIND
Paul Scherrer Institut,
Villigen, Switzerland

A. KHAMAZA
Scientific and Engineering Centre for Nuclear and Radiation Safety,
Moscow, Russian Federation

O. PECHERYTSIA
State Scientific and Technical Center for Nuclear and Radiation Safety,
Kiev, Ukraine

M. HREHOR
Ustav Jaderneho Vyzkumu Rez A. S.,
Hsinec-Rez, Czech Republic

E.-K. PUSKA
VTT Technical Research Centre of Finland,
Espoo, Finland

P. LISKA
Nuclear Power Plant Research Institute Trnava,
Trnava, Slovakia

Abstract

ETSON is the network of ten major European Technical Safety Organizations (TSOs) and of three associated TSOs from Japan, Ukraine and the Russian Federation. ETSON aims at the convergence of nuclear safety practices in Europe by exchanging on nuclear safety assessment guidelines and by collaboration in research. As regards the Fukushima Daiichi NPP accident, the ETSON members obtained deep insights into the course of the accident including related human factor and emergency management aspects. With its knowledge about gaps in the understanding of safety relevant phenomena and about the needs for safety improvements, ETSON is an important driver for the definition and conduction of common post Fukushima research activities. Still in 2011, ETSON presented a research and development position paper that identifies the main research topics also taking into account the lessons learned from the Fukushima accident. Since then ETSON has been spending continuous efforts to further prioritize the identified topics and to define coordinated research projects. In spring 2014, the

ETSON Research Group held a workshop to exchange results of ongoing projects and to share views about common future activities. The workshop focused on the improved simulation of the Fukushima accident, including core degradation, vessel failure, and ex-vessel phenomena as well as Hydrogen distribution and explosion. Among others, it also highlighted the efforts to better understand the phenomena governing potential accident progression in spent fuel pools, and e.g. to improve the capability for fast and reliable source term assessment. Common work is also directed towards the support to IRSN in the development of the European severe accident reference code ASTEC. In order to efficiently work on these priorities, the ETSON members participate in research projects of OECD/NEA like BSAF ("Benchmark Study of the Accident at the Fukushima Daiichi NPP") and they collaborate in the framework of EURATOM projects like CESAM ("Code for European Severe Accident Management"). The paper will provide an overview on the Fukushima related research pursued by ETSON members in the framework of international research programs.

OFF-SITE POST-ACCIDENT RECOVERY AFTER THE FUKUSHIMA DAIICHI ACCIDENT: CHALLENGES AND SOLUTIONS

B. J. HOWARD
Centre for Ecology and Hydrology,
Lancaster, United Kingdom
Email: bjho@ceh.ac.uk

Abstract

Remediation after the Fukushima accident, which has largely focused on the reduction of external dose rates, is being carried out over a wide area, including the municipalities which were evacuated. Selected challenges faced during remediation at Fukushima are described. Technical and scientific support which can enhance national and international capabilities to carry out remediation in the post-accident recovery phase are discussed. The text draws on the outcomes of IAEA activities such as a recent report on the Follow–up International Mission on remediation of large contaminated areas off-site the Fukushima Daiichi Nuclear power plant and the International Experts Meeting on Decommissioning and Remediation.

INTERFACE ISSUES
(TOPICAL SESSION 2)

Chairperson

S. WEST
United States of America

The second topical session presented the wide array of challenges and issues —
including safety and security issues — that TSOs face when interacting with the
regulatory body, the nuclear industry and the public during emergency and non-
emergency situations. In particular, it was stressed that further work is needed
regarding management of possible conflicts of interest when TSOs work for both
operators and regulatory bodies, and that an appropriate mechanism is required for
communication of best practices and feedback related to nuclear security issues. The
potential benefit of TSOs expanding their mandate to address security alongside
safety issues was also raised.

SAFETY AND SECURITY INTERFACES IN EMERGENCY SITUATIONS

J. Jalouneix
Institute for Radiological Protection and Nuclear Safety (IRSN)
France

Abstract

The presentation will focus on table top exercises which create an easy forum to merge safety and security issues and to allow training and open discussions between stakeholders on key points. The general statements and remarks resulting from the French experience in nuclear security exercises will be shared and detailed in 4 main topics : 1 the decision making process, 2 the coordination and interfaces, 3 the planning, preparation and training, 4 the time and people management.

MEETING THE CHALLENGE OF THE SAFETY-SECURITY INTERFACE: IAEA'S ROLE IN SUPPORTING THE ENHANCEMENT OF TECHNICAL COMPETENCE AND SUPPORT FOR NUCLEAR SECURITY WITHIN TECHNICAL SUPPORT ORGANIZATIONS

R. EVANS
Department of Nuclear Safety and Security,
International Atomic Energy Agency,
Vienna, Austria

Abstract

Nuclear security and nuclear safety have in common the aim of protecting persons, property, society and the environment, in the case of safety from the harmful consequences of ionizing radiation and in the case of nuclear security from the harmful consequences of a nuclear security event. Establishment of effective nuclear security measures require an understanding of the interface between safety and security measures and an awareness of the need to optimize the effectiveness of each. Nuclear security requires a State to focus on prevention of, detection of and response to criminal and intentional unauthorized acts directed at or involving nuclear material, radioactive material, associated facilities and associated activities. Nuclear security is a State responsibility and developing and implementing an effective national nuclear security infrastructure is a key requirement for every country and is built on a foundation of legal, regulatory, technical and administrative competence in nuclear security. Appropriate management of the interface between safety and security results in both safety and security in a State being strengthened and enhances each State's capacity to protect and secure its nuclear and other radioactive material, associated facilities and associated activities. Traditionally many Technical Support Organizations have been focused on support for nuclear safety. However, in many countries this has changed over time, particularly where a single regulatory body has responsibility for safety and security and has access to the services of a Technical Support Organization. The last international conference on Technical Support Organizations convened in Tokyo recommended that Technical Support Organization functions be extended to providing technical support to competent authorities in the field of nuclear security. This paper will focus on the role of the IAEA in supporting, upon request, the development of technical competence in nuclear security in all States in order that the safety-security interface is appropriately managed and the capacities of technical support organizations be expanded to include nuclear security. This paper will examine a number of modalities for this support, including national and regional nuclear security support centres, collaborative knowledge networks, guidance and training.

HOW TO MEET THE CHALLENGES IN PUBLIC COMMUNICATION: KINS EXPERIENCES AND PRACTICES

Y. HAH
Korea Institute of Nuclear Safety,
Korea, Republic of

Abstract

This paper presents the main challenges faced by TSOs in communicating with the public. The Fukushima Daiichi Accident underlined the importance of effective communication during crisis situations and as well drew more attention to the need of improving day-to-day communication approaches. As all nuclear related organizations have been confronted with the same challenges, TSOs have also been requested to promote good communication environments for the public in which clear, consistent, and technically sound messages are continuously provided to ensure that the public is well protected against any possible nuclear related problems. The first part of the paper will address the same challenges shared by regulators and TSOs in communicating with the public. It shall reflect the general working conditions where both regulator and TSO are situated in interfacing with the public and stakeholders. After identifying the key elements in public communication that can be applied to any communication strategies, understanding the respective roles of regulators and TSOs will be presented. In general, nuclear regulators are the main contacts for any questions and information regarding nuclear safety and regulation while TSOs are more frequently requested to provide some sort of technical explanation and scientific background about the regulator's decisions. From the coherent communication perspective, both of them need to work together effectively as to provide a synergy, and to avoid any possible conflicts. A global approach is also discussed to be more proactive in communicating with the international public - we are living in a globalized society and are confronting the same challenges to address the public's concern in a more effective and coherent way. Since the TSOs need to be adequately trusted by the public for its technical competence, the need to cooperate among TSOs for effective communication is also highlighted. As part of the good communication practices, Korea Institute of Nuclear Safety (KINS) activities regarding public communication and awareness will be shared. As a unique TSO, KINS has greatly contributed to enhance the public's understanding about its nuclear regulatory activities by supporting the Nuclear Safety and Security Commission (NSSC, Korean regulatory authority) and also carrying out its own outreach activities to cover various stakeholders and public.

NUCLEAR SAFETY AND NUCLEAR SECURITY

J. BYTTEBIER
World Association of Nuclear Operator,
London, United Kingdom

Abstract

In the presentation of Jo Byttebier, the operating experience programme director of WANO will discuss the topic of nuclear safety versus nuclear security. The World Association of Nuclear Operators is a membership organization for operational nuclear power plants and reprocessing plants that was formed in 1989 as a result of Chernobyl accident. The challenges that both nuclear safety and nuclear security are facing will be looked at and the methodology WANO is using to improve and maximize the safety and reliability of the nuclear power plants worldwide will be discussed. From a personal point of view, he will indicate which main directions he would expect nuclear power plants to take to enhance nuclear security in their plants.

EMERGENCY PREPAREDNESS AND RESPONSE
(TOPICAL SESSION 3)

Chairperson

L. BOLSHOV
Russian Federation

The third technical session focused on the roles of and challenges faced by TSOs in terms of emergency preparedness and response; assessment, prognosis and monitoring of emergency situations; and the regulatory and legislative frameworks in place in some countries to protect TSO experts during an emergency response. It was stressed that planning and preparedness are fundamental to the capability to implement effective response actions in nuclear and radiological emergencies. Participants also highlighted the IAEA safety standards and requirements and the IAEA nuclear security guidance as a solid base for nuclear and radiological emergency planning and preparedness. TSOs play an essential role in emergency preparedness and response by providing expertise for governments and for national and international partners. It was suggested that the IAEA's Response Assistance Network (RANET) mechanism could be employed to provide technical expertise, on request, in response to a nuclear or radiological emergency. The involvement of TSOs in providing technical expertise through the IAEA RANET mechanism could support the implementation of the IAEA Action Plan on Nuclear Safety with regard to the assessment and prognosis process. To do so, practical arrangements would have to be established with regard to the process to be followed by registered National Assistance Capabilities within RANET. It was suggested that TSOs should cooperate to develop harmonized approaches to the nuclear safety and security analysis employed in nuclear and radiological emergencies, based on results of benchmarking of tools and methodologies. Participants also pointed to the need to strengthen emergency preparedness and response activities in response to lessons learned from recent nuclear accidents, and the need for emergency preparedness to address security and the safety aspects in a coordinated way.

ROLE OF KINS FOR EMERGENCY PREPAREDNESS AND RESPONSE IN KOREA

S-Y. JEONG
Korea Institute of Nuclear Safety,
Korea, Republic of

Abstract

This paper reviews the important role of Korea Institute of Nuclear Safety (KINS) for emergency preparedness and response in Korea. The KINS performs the regulation for the safety performance and radiological emergency preparedness of the nuclear facilities and radiation utilizations. The radiological emergency preparedness in Korea is based on the Act on Physical Protection and Radiological Emergency which stipulate a national preparation against radiological emergency. Also, KINS has set up the "Radiological Emergency Technical Advisory Plan" and the associated procedures such as an emergency response manual in consideration of the IAEA Safety Standards GS-R-2 and GS-G-2.1. The Radiological Emergency Technical Advisory Center (RETAC), which is in charge of providing technical advice on radiological emergency response, dispatching technical advisory teams to the affected Off-site Emergency Management Center (OEMC), initiating emergency operation of 128 nation-wide environmental radioactivity monitoring stations in accordance with the Nationwide Environmental Radioactivity Monitoring Plan, coordination and control of off-site radiation monitoring, offering radiation monitoring cars, and monitoring the response activities of the operator, will be organized by KINS for the response of emergency situations. Moreover, so as to efficiently implement technical support activities for protection of the public and the environment in a nuclear or radiological emergency of a nuclear power plant, the "Atomic Computerized Technical Advisory System for a Radiological Emergency" (AtomCARE) has been developed and is in operation. Through the system, any nuclear or radiological emergency and its consequences can be quickly verified and assessed, and subsequently, comprehensive management of the information related to public protective actions is also made possible. Recently, The KINS published the report (2013) to adapt the Precautionary Action Zone (PAZ) and Urgent Protective Action Planning Zone (UPZ) applying the IAEA guidelines and reflecting the lessons learned from Fukushima Daiichi NPP accident and was legislated in the Radiological Emergency Act.

POTENTIAL ROLE OF TSOS IN IAEA'S ASSESSMENT AND PROGNOSIS IN RESPONSE TO AN EMERGENCY AT A NUCLEAR POWER PLANT[1]

F. BACIU
International Atomic Energy Agency
Vienna, Austria

---

[1] Although a presentation was given, no abstract or paper was made available. The author's PowerPoint presentation appears on the CD-ROM accompanying this book.

IRSN ACTIVITIES AND EXPERIENCES IN EMERGENCY PREPAREDNESS AND RESPONSE

O. ISNARD
Institute for Radiological Protection and Nuclear Security,
Fontenay-aux-Roses, France
Email: olivier.isnard@irsn.fr

Abstract

After the nuclear accident at Fukushima Daiichi Nuclear Power Plant, Japan in 2011, France decided to enhance its operational capability to respond to any nuclear and radiological emergency, at the governmental level, and to take into account acquired experience during the 2011 response. In this context, the French Government has developed a National Response Plan to a Major Nuclear or Radiological Accident. This plan, addresses the actions to be taken in the areas of protection of the population, but also the aspects of strategic communication or post-accidental management. The French Institute for Radiological Protection and Nuclear Safety, IRSN, fits easily into this governmental scheme, as the national expert in radiological and nuclear risk to public authorities. IRSN is independent of the French operators and as developed specific operational technical capabilities, supported by its area of expertise. IRSN developed its own response organization to meet governmental needs especially in the expertise area. IRSN provides to the French government on an operational basis an expertise capability in the safety and in the accident progression field, in the assessment of radiological consequences field through the evaluation of doses to the public, the monitoring of the environment and also the public. The current paper presents the main features of the capabilities of IRSN in the framework of expertise needed during any response to a nuclear or radiological emergency. The organization, methodologies and tools developed by IRSN to fulfil its duty is also presented.

CANADIAN TSO EXPERIENCE DURING MAJOR NATIONAL EXERCISE (UNIFIED RESPONSE)

G. FRAPPIER
C. COLE
Canadian Nuclear Safety Commission,
Canada

Abstract

The role of the Canadian Nuclear Safety Commission (CNSC) during a nuclear emergency is to provide assurance that appropriate actions are taken by the licensee and response organizations to limit the risk to health, safety, security of the public and the environment. This includes an independent assessment of the onsite conditions and potential offsite consequences as well as assessing and confirming both the licensee's and the responsible government's recommendations concerning protective measures. The Technical Support Branch (TSB) of the CNSC is the integrated TSO of the Canadian nuclear regulator and is responsible for conducting this independent assessment. In May 2014, Exercise Unified Response (ExUR), a three day national level emergency preparedness exercise, was conducted at Darlington Nuclear Generating Station (DNGS). More than 50 government agencies and regional organizations, including the Government of Canada, the CNSC, the Government of Ontario, Ontario Power Generation, the Regional Municipality of Durham and the Municipality of Clarington worked together to test and validate emergency response plans and processes to demonstrate Canada's collective ability to respond to a nuclear emergency. ExUR was a full scale severe accident emergency based on a single-unit loss-of-coolant accident followed by a tornado-initiated full station (4 unit) blackout. Day 3 of the exercise included a simulated radiological release in order to include participation from local authorities in carrying out protective measures for the public. The CNSC fully activated its Emergency Operations Centre (EOC) for the duration of the exercise. Select staff from TSB formed the Technical Assessment Section which provided round the clock assessments of the accident progression, potential source term estimates, and subsequent dispersion and dose evaluations. The team established communication links with international players including the US NRC and the IAEA in order to share technical information and plant status updates. The exercise highlighted many positive aspects of the CNSC response, but also identified areas for improvement. On a positive note the Technical Assessment Section's response indicated that it has a clear role and is able to carry out this role with well established procedures. However, the section was limited in its capability due to the limited plant data available. As well, with the large number of national and international players, the continuous requests for information showed that the section was undermanned. The CNSC has noted all lessons learned and is committed to take the necessary steps to improve its technical response capability. An EOC Improvement Team has been established and will be focusing its efforts in five key areas: (1) Re-assessing the EOC venue; (2) Reviewing current manning levels so that all national and international obligations are met; (3) Improving plant data transfer during accidents from the licensee to the regulator; (4) Implementing a program to develop a state-of-the-art accident assessment tool package; and (5) Developing a comprehensive e-library repository of nuclear power plant information. Implementation of the EOC Improvement Project will ensure that the CNSC Technical Assessment Section's processes and capabilities are in line with best international practices and allow the CNSC to fully meet its mandate during a nuclear emergency.

LEGISLATIVE AND REGULATORY FRAMEWORK FOR PROTECTING EMERGENCY WORKERS IN UKRAINE

S. CHUPRYNA
I. SHEVCHENKO
V. BOGORAD
T. LITVINSKAYA
State Scientific and Technical Centre for Nuclear and Radiation Safety,
Ukraine

Abstract

Issues related to protection of emergency workers are regulated in Ukraine by a number of regulatory documents. Among them, there are documents of 1) upper legislative level, such as the Code of Civil Protection of Ukraine, the Law of Ukraine "On Human Protection against Ionising Radiation"; 2) state safety standards, such as Radiation Safety Standards of Ukraine (NRBU-97) and Basic Health and Safety Rules of Ukraine (OSPU-2005); 3) normative documents of the regulatory authority, such as "General Safety Provisions for NPPs", "Requirements for NPP On-Site and Off-Site Emergency Centres"; 4) normative documents of the operating organisation, such as "Standard Emergency Plan for NPPs of Ukraine", Procedure on Planning Doses of Emergency Workers, Procedure on Conducting Individual Dosimetry Control of External and Internal Exposure of Personnel in Conditions of Emergency, etc. The paper will present information on health and radiation regulations, procedures on issuing permission for higher exposure, definition of emergency personnel, and measures on protecting emergency workers. The main technical and organisational measures on protecting emergency workers include supervision over non-exceeding health and radiation regulations, restriction of exposure, conducting radiation survey on NPP premises and site, prophylaxis of external and internal exposure of personnel, decontamination, medical protection, arrangements for continuous monitoring and recording of doses received by emergency workers, procedures to ensure that doses received and contamination are monitored in accordance with established guidance and international standards, and arrangements for the provision of appropriate specialized protective equipment, both individual and collective, procedures and training for emergency response in the anticipated hazardous conditions, etc. Conclusions will be given on conformity of the Ukrainian regulatory framework in force for protecting emergency workers with the IAEA Safety Requirements GS-R-2 "Preparedness and Response for a Nuclear or Radiological Emergency".

ROLE OF THE TECHNICAL EXPERTS IN THE CONVEX 3

I. SOUFI
National Center for Nuclear Energy, Sciences and Techniques,
Morocco

Abstract

Morocco hosted on 20-21 November 2013 a ConvEx 3 Exercise, co-named Bab Al Maghrib. This type of exercise is a large scale joint international exercise covering the early phase of a severe radiation emergency, based on a national exercise conducted in Morocco. The purpose of the ConvEx-3 exercises is to evaluate response to a major radiation emergency and, in particular, to evaluate the exchange of information, provision of the international assistance and coordination of public information. The 2013 CONVEX 3 exercise was based on a severe radiological emergency triggered by nuclear security events with transnational/transboundary implications. The exercise was conducted for 25 hours on real time, with the participation of 59 Member State and 10 International Organizations. The exercise scenario was based on evolving threats and multiple nuclear security events (RDDs) in populated places. The following issues were addressed in the exercise scenario: (a) dispersion of radioactive material into the atmosphere, (b) the interface between safety and security, (c) medical and public health, (d) impact on commerce, industry and tourism (food and products contamination, contamination of vehicles, ships), and (e) communication with the public. Additionally, at the international level, the exercise focused on different protective and other response actions in connection with activities such as commerce, industry and tourism. Protecting the public, the responders and mitigating the consequences require a clear view and understanding of the situation, the risks and the potential consequences. With this regard, beside the national organizations and authorities in charge of lifesaving, security, public protection, control command and coordination, several technical organizations (such as meteorology, aviation, maritime, health, regulatory body and TSO organizations) played an important role in assisting national authorities responsible for decision making and the protection of the public to have a better understanding of the risks and their potential consequences. Around 15 national organizations participated to the exercise playing in three activated emergency centers: On Scene (managing the situation on the scene), at the National Emergency Center / NEC (ensuring decision making, coordination and allocation of resources), at the Technical Crisis Center. Throughout the exercise preparation and conduct, the technical experts provide a substantial support: technical injects, exercise documents, meteorological data, plume modeling, on scene radiation safety, radiological search and survey, radiological environmental survey, decontamination, etc. In particular, the Technical Crisis Center (TCC) located at the National Center for Nuclear Energy, Sciences and Techniques (CNESTEN) acting as TSO supported the National Emergency Center (NEC) for decision making regarding the protective actions. The main mission of the TCC was to support the NEC by performing radiological risk assessment and making recommendations on protective actions for the public and the emergency responders. The presentation will focus on the presentation of the exercise features and the role and involvement of the technical experts to support both the exercise preparation and conduct and the response to such events.

NUCLEAR AND RADIATION ACCIDENT EMERGENCY RESPONE AND RADIATION ENVIRONMENTAL MONITORING IN CHINA

H. LIU
National Nuclear Safety Administration (NNSA),
Beijing, China

Abstract

This paper introduced the regulatory framework, legal system and organizational structure of the nuclear and radiation accident emergency preparedness and response as well as radiation environmental monitoring in China. It described the roles and responsibilities of the National Nuclear Safety Administration and its TSOs in this regard and their major activities. In addition, the paper introduced the improvements made by the government in the field of nuclear and radiation accident emergency preparedness and response after the Fukushima nuclear accident, the latest plan and perspectives regarding nuclear and radiation emergency response in China, and the challenges and planned improvements for NNSA/NSC in this field.

MAINTAINING AND STRENGTHENING TSO CAPABILITIES
(TOPICAL SESSION 4)

Chairperson

A. DELA ROSA
Philippines

The fourth technical session presented the challenges faced by TSOs in maintaining professional expertise and building capacity, and in understanding the human and organizational factors that can affect both. It was recognized that incorporating such factors into the regulatory framework to improve safety is a good practice and can be accomplished using a systematic and formalized approach. During the session, it was stressed that the management system applied within a TSO is essential, and that the use of knowledge management tools such as the IAEA's Guidelines for Systematic Assessment of Regulatory Competence Needs (SARCoN) can support the assessment and management of the required knowledge and address future gaps and needs. A structured and robust training programme is needed to build and sustain the critical skills required of a TSO. Technical support is essential for safety of nuclear power plants and, in decision making organizations, for achieving sound and effective decisions on their safety. It was noted that each organization needs to decide on the best organizational structure based on its organizational factors, including but not limited to corporate philosophy, strategy, task at hand, and human and financial resources. In this decision, the choice and extent of internal versus external TSO use needs to be carefully considered to ensure the adequacy, effectiveness, quality and timeliness of the technical support to be received. It was noted that provision of resources and strong government and industrial support enable TSOs to enhance their support of the safe use of nuclear power. It was suggested that a consensus on international standards describing the establishment and maintenance of effective technical support would enable countries with nuclear power programmes to make better use of technical support.

HUMAN AND ORGAINZATIONAL FACTORS

52

K. HEPPELL-MASYS
Canadian Nuclear Safety Commission,
Ottawa, Canada

Abstract

The integration of Human and Organizational Factors (HOF) within a regulatory framework will strengthen the actions of a Technical Support Organization (TSO), lead to a more effective regulatory oversight and result in improved nuclear safety regulatory system. The importance of HOF has long been recognized as critical to safe operations. As safety results from the interaction of individuals with technology within the organization, as indicated in the IAEA in Safety Standard GS-G-3.5, "The Management System for Nuclear Installations", a sound safety oversight should encompass this interaction as well." This presentation will describe how the Canadian Nuclear Safety Commission (CNSC) has developed a robust regulatory framework which supports our oversight in the area of HOF. CNSC's Safety and Control Area framework explicitly identifies the integration of HOF within its regulatory oversight activities. While there is still work to be done, practical examples are provided which demonstrate how the CNSC has achieved successful integration amongst technical disciplines and the benefits realized from this approach. One of the most significant benefits is in the synergy created when specialists from various disciplines interact, share knowledge and approach safety from a holistic perspective. This integrated approach ensures the continuous development and availability of the scientific expertise necessary to support an effective nuclear safety regulatory system.

BRIDGING REQUIRED CAPABILITIES AND TRAINING

P. DE GELDER
B. BERNARD
P.MIGNOT
M. ROOBAERT
Bel V,
Brussels, Belgium

Abstract

For Bel V, the Belgian TSO, developing and maintaining required competence and expertise is of utmost importance. Essential activities to achieve this goal are embedded within several processes of the Bel V integrated management system (IMS) which is certified according to ISO 9001:2008. The main processes of interest are:

- Deliver expert services in nuclear safety and radiation protection;
- Manage expertise and technical quality;
- Manage Human Resources.

The presentation will explain the role of Bel V's Technical Responsibility Centres (TRC) that play a key role in the process on "Deliver expert services in nuclear safety and radiation protection", especially for the identification of required expertise, and for daily managing and periodic evaluation of the expertise. The process "Manage expertise and technical quality" covers aspects related to knowledge management. Examples of available tools will be presented that are important for evaluating the vulnerability of our expertise (by using the in house developed Knowledge Critical Grid) or for ensuring knowledge transfer where needed. Further, the process "Manage Human Resources" plays an important role by covering all aspects related to recruitment, by providing role descriptions, and by describing how to manage competence and training. Within that process, an important effort was recently started in view of a more structured identification of the individual existing competence (KSA) gaps using the SARCoN tool, on the basis of a reference list of KSAs and the role descriptions. Our first findings after having performed a pilot-project for implementation will be presented. Besides the investments on developing and maintaining technical expertise, Bel V also decided recently to launch an important effort on non-technical training through participation of the whole staff in an "Interpersonal Effectiveness Development Path", coordinated by an external consultant company. The main aspects of this effort will be presented.

MEETING CHALLENGES OF PROFESSIONAL DEVELOPMENT OF EU TECHNICAL SAFETY ORGANISATIONS EXPERTS

D. LOUVAT
European Nuclear Safety Training & Tutoring Institute,
Fontenay aux Roses, France

Abstract

The support provided by Technical Safety Organizations (TSOs) to Nuclear Regulatory Authorities (NRAs) in carrying out their designated functions, depends on highly qualified personnel who are competent in many disciplines. The development and maintenance of this workforce needs on-going attention from governments and stakeholders to ensure that adequately skilled and competent personnel are available at any time, taking into consideration retirements and the continuous need for personnel resulting from natural fluctuation, from new developments or national requirements. In the European Union, this demand for skilled personnel set against a generally ageing workforce makes it very clear that it is high time to put in place a training mechanism that ensures the maintenance of the current skilled and competent personnel at NRAs and TSOs, and the flow of new recruits for long-term sustainability. In the light of the above identified need, and in the aftermath of the Fukushima Daiichi accident, the European Commission took action and launched a project for "Sharing & Growing Nuclear Safety Competences" (NUSHARE project). This project aims at strengthening nuclear safety and fostering a common nuclear safety culture in the EU-28. One out of three working packages of the NUSHARE project is dedicated to the development of a comprehensive training programme for new entrants, professional staff already working at NRAs or TSOs, or experts who wish to start a career in this field. This important task is coordinated by the European Nuclear Safety Training and Tutoring Institute (ENSTTI), an initiative of the European Technical Safety Organizations Network-ETSON. ENSTTI is a centre specialized in meeting the growing need for highly qualified personnel with adequate knowledge and skills in nuclear safety and security at NRAs and TSOs. The paper provides a brief overview of the NUSHARE project with a focus on the development of a comprehensive training programme tailored to the requirements of NRAs and TSOs. In addition, the paper highlights one module of the training programme that is dedicated to the systematic development of entry-level skills necessary for employment at NRAs or TSOs. ENSTTI addresses also the above indicated issues by developing a comprehensive training catalogue implementing the European Credit system for Vocational Education and Training (ECVET) which is already used in many other industrial sectors. The objective of the ECVET system is to promote mutual trust, transparency and mutual recognition of acquired learning outcomes in the form of ECVET credits.

POST FUKUSHIMA ACTIVITIES AT AECL

S.J. BUSHBY
Atomic Energy of Canada Ltd,
Chalk River, Canada

Abstract

Atomic Energy of Canada Limited (AECL) is Canada's premier nuclear Science and Technology (S) organization. AECL's capabilities have been utilized extensively to provide technical support to government and industry partners as the events of Fukushima unfolded. They continue to play a role in supporting longer-term S for both the industry and the regulator to demonstrate defense in depth for Canadian nuclear facilities.

A STRONG AND VIABLE TECHNICAL SERVICE ORGANIZATION TO MEET CURRENT AND FUTURE REGULATORY CHALLENGES – NRC'S VISION AND PERSPECTIVES

K. S. WEST
B. THOMAS
U.S. Nuclear Regulatory Commission,
Washington D.C., United States of America

Abstract

The Office of Nuclear Regulatory Research (RES) is, as established under Statute, a US-NRC program office that develops and maintains technical tools, analytical models, analyses, experimental data, and technical guidance needed to support the agency's regulatory decisions. RES is essentially the NRC's statutorily mandated technical support organization (TSO) that provides technical expertise and capabilities to support NRC's program offices, namely the Office of Nuclear reactor regulation (NRR), the Office of New Reactors (NRO), the Office of Nuclear Material Safety and Safeguards (NMSS), and the Office of Nuclear Safety and Incident Response (NSIR) in licensing and regulatory decisions. RES develops the technical bases to confirm that the methods and data generated by the nuclear industry help ensure that adequate safety isestablished and maintained. In addition to conducting confirmatory research, as a technical support organization, RES conducts anticipatory research whereby it develops expertise and capabilities to evaluate longer term (approximately five years and beyond) needs of the Agency. To provide the technical bases for future regulatory decisions, RES looks where the regulated industry is moving and conducts exploratory research as needed to prepare the USNRC to respond to industry requests and initiatives. The paper will provide some examples of the technical activities and support provided by RES in support of US-NRC mission. The core capabilities required to continuously provide these technical services are of paramount importance to RES. In addition to regulating the commercial use of radioactive materials to protect public health and safety and to protect the environment, the USNRC has responsibility for protecting and safeguarding nuclear materials and nuclear power plants in the interest of national security. Hence, RES also provides research and technical support to broad government-wide initiatives associated with national security. Thus, safety and security culture is an integral part of RES and US-NRC. In its broadest sense, "safety culture" refers to how well the NRC's mission, policies, and working environment support nuclear safety and security as the agency's overriding priorities. RES ensures that personnel in the safety and security sectors have an appreciation for the importance of each, emphasizing the need for integration and balance to achieve both safety and security in their activities. It is important that consideration of these activities be integrated so as not to diminish or adversely affect either; thus, mechanisms should be established to identify and resolve these differences. To this end, several important programs, such as Open and Collaborative Work Environment, have been put in place. RES's principal product is knowledge; thus, knowledge management (KM) is an integral part of the RES mission. RES's objective is to capture, preserve, and transfer key knowledge among employees and stakeholders. The body of knowledge can be used when making regulatory and policy decisions and ensures that issues are viewed and analyzed within a historical context. RES KM activities include participation in an Agency-Level KM Steering Committee to help to promote and cultivate an awareness of the value of KM, expansion of Expertise Exchange Program, continuation of support to communities of practice (CoPs), and championing KM development and preservation.

ORGANIZATION AND RESPONSABILITIES OF INTERNAL OR EXTERNAL TSOs

N. KILIC
International Atomic Energy Agency,
Vienna, Austria

Abstract

It has been recognized that the provision of effective technical support is essential to optimize the safe operation of nuclear power plants and to maximize their reliability, availability and productivity. A competent TSO can greatly support the nuclear programme leaders, the appointed government organizations and other stakeholders to provide all necessary information to support their decisions including during the licensing, construction, commissioning and operation phases. The importance of a Technical Support Organizations (TSO) was enhanced again in the technical reviews and activities that followed the Fukushima accident. In the light of collected operational experience since the issue of IAEA's Roles and Responsibilities of Technical Support Organizations (TSO) document (TECDOC-1078), IAEA initiated a revision and update to such guidance in mid-2012 for improvement of TSO organizations and their functions. The new guidance is also to serve as a reference for new comer countries, where establishing an effective TSO both in operating organization and the regulatory body becomes essential for successful and safe implementation of the first nuclear power plant. This paper discusses the IAEA activities on the guidance and provides key points, such as organization, roles and functions, which have been elaborated during the consultancy and technical meetings with Member States experts.

THE OBJECTIVES AND CURRENT PROGRESS TO STRENGTHEN CAPABILITY BUILDING IN CHINA

J. CHENG
National Nuclear Safety Administration,
Beijing, China

Abstract

TSOs have been playing very important roles during the development of nuclear energy in China. This paper describes TSO's values, functions capability requirements and general situation, introduces the target of strengthen TSO capability building in China as well as current achievements and progress, including establishment and construction of China's National Research and development Base on Nuclear and Radiation Safety.

NETWORKING AMONG TSOs AND BEYOND
(TOPICAL SESSION 5)

Chairperson

H. KIM
Republic of Korea

The fifth technical session presented the issues and challenges that TSOs face, as well as the benefits they derive, when networking in a global environment. The papers presented in this session indicated that both TSOs and regulators around the world are very actively involved in networking. These bodies share the goal of improving nuclear and radiation safety while at the same time optimizing resources and harmonizing technical and scientific support, particularly to provide independent, objective and technically strong evaluations.

The presentations focused on the considerable networking taking place among TSOs and regulatory bodies, especially within Europe and among countries with more advanced nuclear power programmes. Participants noted that, with the potential mutual benefits that could be obtained from sharing knowledge and experience, exploring synergies in a wider international context may be of interest to many countries, especially for those countries in Asia, Africa and Latin America and the Caribbean that do not have nuclear power and yet could greatly benefit from the scientific and technical expertise that is available from the TSOs represented at the conference. This could, for example, be facilitated through the IAEA's TSO Forum, which could be used to reach out to both nuclear and non-nuclear countries. Looking to future networking opportunities, TSOs may consider expanding R&D activities, for example, to encompass decommissioning, remediation, human and organizational factors, safety analysis and predications in the case of emergencies, as well as further research on defence in depth, to minimize potential consequences to the public and the environment from severe accidents. It was also noted that engaging the younger generation in nuclear and radiation safety is key to the sustainability of future expertise. As such, activities such as those being carried out in this area by ETSON are to be further encouraged.

THE TSO FORUM IN THE GLOBAL NUCLEAR SAFETY AND SECURITY NETWORK (GNSSN)

M. HEITSCH
European Commission,
Joint Research Centre,
Petten, Netherlands

L. GUO
International Atomic Energy Agency,
Vienna, Austria

Abstract

The Technical and Scientific Support organization (TSO) Forum also known as TSOF was established following the recommendations of the TSO Conference 2010 in Tokyo. The objectives of the Forum as defined in the Terms-of-Reference are to encourage open dialogue and sharing of scientific and technical information among the technical experts and their TSOs in general worldwide. The Forum is open to all member States after official nomination. The IAEA hosts numerous networks to support the international efforts to enhance nuclear safety and security. The Global Nuclear Safety and Security Network (GNSSN) integrates numerous global, regional and thematic networks, is a working platform and document archive and gives access to many information resources of the IAEA. The TSO Forum is fully integrated in GNSSN. The paper gives an overview of the structure of TSOF in GNSSN. Like many other networks it includes a public and a password protected section. Major activities of the Forum and their reflection in the network are discussed. TSOF has a Steering Committee with chair and vice-chair. Major meetings at IAEA and internationally are used to promote the Forum. Those technical experts who are not yet members of the Forum will be invited to join TSOF and to benefit from enhanced international collaboration.

NETWORKING ACTIVITIES IN ETSON

M. HREHOR
Nuclear Research Institute Rez,
Rez, Czech Republic

Abstract

Nuclear Safety Convention, EU Council Directive 2014/87/Euratom of 8 July 2014 amending Directive 2009/71/Euratom establishing a Community framework for the nuclear safety of nuclear installations and a number of publications of the IAEA and OECD Nuclear Energy Agency declare the vital role of the state, even in today's market conditions, in the field of peaceful use of nuclear energy, in particular in creating the political, legal and regulatory framework and long-term strategy with a focus on sustainability and public awareness. The key is in particular the regulatory role of the state in the field of nuclear safety and the necessary expertise in this area. In a number of countries operating nuclear power plants there exist independent scientific expert organizations, so called "Technical Safety Organizations - TSO", providing a comprehensive view of the long-term safety of nuclear installations and maintaining a high level of expertise on the basis of research and development activities. The paper describes the basic mission of TSO organizations - support of state regulatory authority in the form of an independent expert, analytical and research services in the field of nuclear safety and radiation protection - and the ways of its implementing. TSOs are developing and maintaining their expertise and skills by carrying out long-term R & D programs aimed, inter alia, at verification of technical solutions proposed by the nuclear industry and at building a comprehensive knowledge base for assessment of nuclear safety. The paper briefly lists the main research topics which are subjects of the TSOs R projects. Nuclear safety assessment requires a high level of proficiency in safety assessment methodologies and in analysis of operating experience feedback as well as. The need for increased co-operation and reinforced sharing of experience in the field of nuclear safety expertise emboldened the competent organizations with nuclear safety expertise in Europe to establish an ETSON association (European Technical Safety Organisation Network) - a network aimed in promoting close cooperation on harmonized approaches to safety issues and their assessments. The paper describes the structure of ETSON association, methods of its working and summarizes the main recent activities and achievements.

LITHUANIA INCREASE NETWORKING AFTER JOINING THE EU

E. USPURAS
Lithuanian Energy Institute,
Lithuania

Abstract

In 1990, after Lithuania declared its independence, the Ignalina NPP came to the jurisdiction of Lithuania, however, all technical scientific support organizations remained in the Russian Federation. Therefore the need to develop the independent institutions of nuclear regulatory and technical support was raised. During the 1991 – 2009 (till the final close of Ignalina NPP) the necessary infrastructure for nuclear regulation and technical-scientific support was created. Lithuanian Energy Institute (LEI) became one of main technical support organizations in Lithuania, capable to perform all necessary safety analyses for Ignalina NPP. The support of Western countries and experience, received participating in different international projects, was very important during the process of experience acquiring. Lithuania's accession to the EU (2004) has opened wide horizons for cooperation. Active participation in European research programmes was very important to Lithuania since it allows Lithuania to further integrate into the European Research Area, benefiting both Lithuanian researchers and country as a whole. Lithuania is actively participating in the European research programmes (FP6, FP7) and this is the basis for successful participation in the biggest EU Research and innovation programme Horizon 2020. However, evaluating the differences in research infrastructures between the "old" and "new" EU Member States, there is only way for new EU states to participate on equal basis with the EU-15 countries – to join forces and work together.

A NEW ENVIRONMENT FOR NUCLEAR SAFETY: MAIN CHALLENGES FOR THE OECD/NEA

J. REIG
OECD Nuclear Energy Agency,
Paris, France

Abstract

The paper presents the new challenges for the NEA after the Fukushima Daiichi accident. All the seven committees of the NEA have initiated specific activities to address the lessons learnt from the accident. The NEA issued a report last year describing the main goals and expectations of these activities as well as an initial response to the accident. (The Fukushima Daiichi Nuclear Power Plant Accident: OECD/NEA Nuclear Safety Response and Lessons Learnt – NEA No. 7161. September 2013). The paper presented at this conference focuses on the activities coordinated by the safety committees, the Committee on Nuclear Regulatory Activities (CNRA) and the Committee on the Safety of Nuclear Installations (CSNI). The CNRA has launched activities related to accident management, defence-in-depth, crisis communication and the safety culture of the regulatory body. The CSNI is looking at different technical issues which played an important role on the evolution of the accident. In addition the CSNI has launched several safety research projects related to the accident scenario and has established a Senior Task Group on Safety Research to achieve a dual objective, to support the Japanese safety institutions and to agree on research activities which will benefit the international community. The paper will conclude describing the current status of the Benchmark Study of the Accident at the Fukushima Daiichi NPP (BSAF), which is completing its phase 1 this November.

FORO, ENHANCING RADIATION PROTECTION AND NUCLEAR SAFETY IN IBERO-AMERICA

F. CASTELLO,
A. DE LOS REYES
Consejo de Seguridad Nuclear,
Madrid, Spain
Email: fcb@csn.es

Abstract

The Ibero-American Forum of Radiological and Nuclear Regulatory Agencies (FORO) was established with the purpose of promoting a high level of radiological and nuclear safety in all practices using radioactive materials in the member countries and, consequently, in the countries of the Ibero-American region. The FORO was created in 1997 and is now composed of the nuclear regulatory bodies of nine countries: Argentina, Brazil, Chile, Colombia, Cuba, Mexico, Peru, Spain and Uruguay. The FORO has the vision of providing a fruitful environment for strengthening the regulatory organizations in its member countries, through the exchange of information, experience and best practices, as well as a robust technical programme in key radiological and nuclear safety areas and regulatory practices identified and supported by the FORO's members. The FORO values cooperation with other organizations while maintaining its independence. The FORO has launched a number of relevant technical projects in close cooperation with the IAEA (FORO's scientific reference), in key areas of nuclear safety and radiological protection. Seven of these projects have been completed: (1) prevention of accidental exposures in radiation therapy through the application of probabilistic risk assessment and the development of a radiotherapy risk assessment tool (named SEVRRA); (2) cooperation between regulatory and health authorities for the regulatory control of medical exposure; (3) regulatory assessment and inspection of ageing management and long term operation in nuclear power plants (NPPs); (4) control of inadvertent radioactive material in scrap metal and recycling industries; (5) assessment of stress tests performed to NPPs in the FORO member countries; (6) preparedness and response to emergencies and (7) licensing and inspection programme for cyclotrons. The results of many of these projects have been published jointly with the IAEA in Spanish –FORO language- and some of them are also available in English. In addition, other activities are being conducted on safety culture in regulatory practices for radioactive sources; the implementation of the radiotherapy risk evaluation tool SEVRRA in FORO member countries; the development of a guide for building, developing and maintaining of competences and training programmes on licensing and inspection of safety of nuclear reactors for regulatory staff; and the clearance of small low level waste from small facilities. The FORO has recently launched new technical projects on implementation of clearance concept and criteria for (small) nuclear installations and on Risk Matrix applied to Industrial Installations. As the results of the technical program become available, dissemination to all countries in the region and beyond as well as their eventual application is basically being carried out through the IAEA's technical cooperation programme in close collaboration with the FORO. The dissemination of these results is achieved through workshops, conferences, technical meetings and the Global Nuclear Safety and Security Network (GNSSN). The FORO collaborative web-based IT platform, the RED, was fully updated and released in June 2013. The new RED aims at further sharing and exchanging knowledge on new or existing nuclear and radiological safety and nuclear security issues of regulatory interest and contributes to the development of national systems for the regulation, authorization and control of activities involving the safe use of ionizing radiation. The RED constitutes a valuable reservoir of technical information on radiation protection, nuclear safety and security from the regulatory perspective. FORO is strengthening its cooperation with other associations, institutions and networks: (1) The World and the Pan-American Health Organizations in the field of radiation protection to patients; (2) The General Secretariat for Ibero-America to facilitate the spreading of the results of its projects; (3) The International Commission on Radiological Protection as Special Liaison Organization to exchange scientific and technical information on radiation safety, (4) The International Radiation Protection Association (IRPA) to cooperate in radiation protection through areas of mutual interest like the safety culture. Around one hundred regulatory specialists in various technical areas related to radiation protection, nuclear safety and security are networking in groups, sharing their experiences, good practices and lessons, addressing problems and regulatory challenges by means of technical projects and other activities. These activities are prioritized by FORO taking into account the members' needs and without duplicating the efforts made by the IAEA in these areas and making the results available to others. FORO, as an association of 9 regulators, provides technical support to its members, to the other regulators in the region, creating a reference in nuclear safety and radiation protection in close cooperation with the IAEA.

THE EUROPEAN CLEARINGHOUSE FOR NPP OPERATING EXPERIENCE FEEDBACK OPERATED BY EC-JRC: NETWORKING EUROPEAN NUCLEAR REGULATORS AND TSOs

M. HEITSCH
M. BIETH
B. ZERGER
European Commission,
Joint Research Centre,
Petten, Netherlands

Abstract

Operating Experience Feedback (OEF) is one of the ways of improving nuclear safety of operating nuclear power plants. Several participants at the conference on Improving Nuclear Safety through Operating Experience Feedback that was held in Germany in 2006 discussed the possibility and benefits of joined efforts at European level to enhance the effectiveness of OEF. As a result a regional initiative has been set up in 2008 in support of EU Member States' nuclear safety regulatory authorities, but also EU technical support organizations, international organizations and the broader nuclear community, to enhance nuclear safety through improvement of the use of lessons learned from operational experience of nuclear power plants (NPPs). The experience of US NRC Operating Experience (OE) Clearinghouse showed that the establishment of a centralized OE Clearinghouse for a particular region in the world can yield significant benefits due to optimized use of resources and improved feedback of lessons learnt. Due to differing regulatory regimes in the EU member countries, significant diversity of the NPP designs and different languages used, the establishment of the European Clearinghouse was more complicated and challenging and needed strong support and commitment from the EU nuclear safety regulatory authorities. The Joint Research Centre (JRC) of the EC has been chosen to play a central role in establishing and running of the European Clearinghouse for OEF. The choice allowed use of well-established JRC working mechanisms, means and technical expertise in the field to promote better cooperation and more effective use of the limited national resources and to strengthen the capabilities for OE analyses and dissemination of the lessons learned. The European Clearinghouse is organized as a Network operated by a Central Office located at the Institute for Energy and Transport that is part of JRC of the European Commission. It gathers 17 European Safety and 3 major European TSOs. The main objectives of the European Clearinghouse are: strengthening co-operation between European Safety Authorities, Technical Support Organizations (TSOs) and the international OEF community to collect, evaluate and share NPP operational events data and apply lessons learnt in a consistent manner throughout member countries; establishment of European best-practice for assessment of operational events in NPPs; coordination of OEF activities and maintenance of effective communication between experts from European regulatory authorities involved in OEF analyses and their TSO; strengthening European resources in operational experience; support for the long-term EU research and policy needs on NPP Operating Experience Feedback. The main activities covered by the European Clearinghouse are: topical studies providing in-depth assessment of preselected subjects related to NPP operating experience; contribution to improve the quality of event reports sent to the International Reporting System; quarterly reports on Operating Experience; development, maintenance and population of a database for storage of Operating Experience related information; and delivery of training in the field of Root Cause Analysis and event investigation. Six years of operation of the European Clearinghouse have shown the added value of the initiative and further areas are being developed such as statistical tools to identify topics on which the efforts should focus in the future.

CLOSING SESSION

The closing session provided an opportunity to discuss future developments and to present a vision of future cooperation among TSOs. The conference participants identified several issues for consideration and proposed that a fourth TSO conference, on TSO functions, science and expertise, could be held in Belgium. The proposal to produce safety guidance describing the performance of TSO functions was also discussed. The Conference President presented the summary and conclusions of the conference.

Among the conference's main conclusions were that peer review missions such as those carried out within the IAEA's Integrated Regulatory Review Service (IRRS) could be implemented to evaluate the capabilities of those national TSO functions that contribute to research, training and education, and the performance of safety assessments. It was also agreed that the work of the TSO Forum could be more widely promoted to provide better support and contribute to the building of the capabilities of newcomer countries. Common nuclear safety research projects could be developed among organizations carrying out TSO functions by using existing frameworks to the extent possible, in particular those provided by the IAEA and the OECD Nuclear Energy Agency, and through other efficient means, such as by joining or establishing regional TSO networks. Strong coordination between TSOs and their authorities could support the registering of their capabilities within the RANET. Further integrating TSO expertise at a high level could enhance the capabilities for assessment and prognosis during a nuclear or radiological emergency. The sustainability and the adequacy of maintaining TSO capacity is a responsibility of individual States. The IAEA was encouraged to consider expanding the activities of the TSO Forum to develop it into a science and expertise forum providing comprehensive coverage of issues concerning TSO functions in nuclear regulation. It was also encouraged to consider establishing, in particular, new means of improving international networking to share knowledge and experience on technical and scientific practices.

SUMMARY AND CONCLUSIONS OF THE CONFERENCE[1]

Benoît DE BOECK
President of the Conference

BACKGROUND TO THE CONFERENCE

The International Conference on Challenges Faced by Technical and Scientific Support Organizations (TSOs) in Enhancing Nuclear Safety and Security, held in Beijing, China, in October 2014, continued the tradition established by the two preceding conferences on this subject that were held, respectively, in Tokyo, Japan, in 2010, and in Aix-en-Provence, France, in 2007. Like these earlier conferences, the outcomes of this conference will play a vital part in the national and international efforts made to ensure the effectiveness of nuclear safety and security regulatory systems and will focus, in particular, on and ways to maintain and strengthen the actions of TSOs in supporting the enhancement of safety and security worldwide.

OBJECTIVES OF THE CONFERENCE

The objectives of this conference were to assess and review ways to further increase the effectiveness of TSOs, taking into account lessons learned from the accident at the Fukushima Daiichi nuclear power plant. In particular, the conference:

• Discussed the role of research and development (R&D) in enhancing nuclear safety;

• Helped participants to understand the impact of the Fukushima Daiichi accident on TSOs and to extract lessons to be learned;

• Highlighted the role of TSOs in the implementation of the IAEA Action Plan on Nuclear Safety;

• Provided a forum for discussion of the roles, functions and value of TSOs in enhancing nuclear and radiation safety and nuclear security, including through capacity building in those countries launching or expanding their nuclear power programmes;

• Facilitated the exchange of experience and good practices in planning and implementing cooperative activities for capacity building and in identifying needs for assistance activities from the standpoint of recipient countries;

• Considered appropriate approaches to enhancing cooperation and effective networking among TSOs and beyond, including the creation of centres of excellence;

• Provided an overview of the technical and scientific support needed to maintain a sustainable nuclear safety and security system;

---

[1] The opinions expressed — and any recommendations made — are those of the participants and do not necessarily represent the views of the IAEA, its Member States or the other cooperating organizations.

•   Fostered continued dialogue on all relevant technical, scientific, organizational and legal aspects at the international level.

OPENING SESSION

Li Ganjie, Vice Minister, Ministry of Environmental Protection, and Administrator of National Nuclear Safety Administration (NNSA), spoke of the phenomenal growth of nuclear power in China and the absolute need for scientific and technical innovation to assist it in developing its nuclear safety infrastructure and nuclear power programme in order to meet its soaring demand for energy, protect the environment, and fulfil its national and international obligations to help mitigate global climate change. To this end, Vice Minister Li Ganjie provided details on the construction of a new national research and development compound for nuclear and radiation safety regulation, reinforcing technical support capacities in nuclear safety review, supervision, monitoring, emergency response, public communication and international cooperation. The phase I building area is 93 000 m^2, with a 750 million RMB investment; the total building area is close to 200 000 m^2. When completed, the compound will serve as a leading international platform for integrated and specialized technical support for nuclear safety regulation.

In addition, last year the funding for TSOs in China grew by 28% and the number of TSOs at the local, regional and national levels grew exponentially. Vice Minister Li Ganjie further stated that "Nuclear safety knows no boundary. Technical exchange shall have no barriers. The TSOs in China have paid full attention to studying and learning from the latest international standards on nuclear safety, and to improving China's nuclear safety regulatory system and legislation and standards." He further envisions that TSOs will soon be shouldered with more challenging work. He proposed that all countries should:

•   Improve their verification, calculation, testing and validation capacity, and build up nuclear safety evaluation and review centres;

•   Make greater efforts in R&D on nuclear safety technologies and build up nuclear safety technology R&D and application centres;

•   Improve nuclear safety information technology and build up data collection and exchange centres;

•   Establish sound human resource nurturing mechanisms and foster human resource 'incubators'.

As NNSA celebrates 30 years of continuous improvement and growth in China, it will release a policy statement on nuclear safety culture describing eight features of nuclear safety culture. The policy statement will include a proposal for creating a sound nuclear safety culture across the entire industry, calling for all countries both to contribute to improving nuclear safety and to put forward their own formal statement on safety culture.

70

Denis Flory, Deputy Director General, Head of the Department of Nuclear Safety and Security, IAEA, detailed the conference objectives, indicating that one of the main challenges facing the nuclear community today is the shortage of specialists with the competencies necessary for ensuring nuclear safety. In addition, he stated that TSOs increasingly provide crucial support for countries expanding or embarking on nuclear power, and that they increasingly work across borders, providing crucial assistance to regulatory bodies in places with fewer resources. He further emphasized that it was necessary to clarify expectations of TSOs in these circumstances, and that countries using an external TSO must have the in-house competencies to review the TSO's work (i.e. they must be 'intelligent customers'). Mr. Flory also underscored that many of the safety assessment actions undertaken after the Fukushima Daiichi accident have relied on close cooperation between the IAEA and TSOs. TSOs are now also involved in nuclear security through the Global Nuclear Safety and Security Network (GNSSN). All stakeholders benefit from the comprehensive information available from TSO networks such as the IAEA TSO Forum and the European Technical Safety Organizations Network (ETSON).

Benoît De Boeck, President of the Conference and General Manager of BEL V, Belgium, stated that TSOs provided valuable support to Governments during the Fukushima Daiichi accident, such as explaining to the public what happened, what could happen and how their own country could be affected; they also recommended improvements to ensure that such an event would be less likely in the future. Mr. De Boeck also emphasized that TSOs provide independent, science based views that enable regulatory bodies to do their work. Therefore, TSOs play an important role in the emerging concept of 'institutional defence in depth'. With regard to protection against extreme events, TSOs contribute to taking a balanced view of assessing risk (external versus internal).

KEYNOTE SPEECHES

Jacques Repussard, Director General of the Institute for Radiological Protection and Nuclear Safety (IRSN), who served as President of the TSO Conference held in 2010 in Tokyo, Japan, provided a progress update on the recommendations made during the 2010 conference. He recalled the first key conclusion of the 2010 Conference, which recognized the two key pillars of a sustainable effective regulatory system: the 'authoritative function' and the scientific functions or 'TSO functions'. He indicated that the progress over the past four years has been mixed: while there has been success in establishing a TSO Forum, organizing a third TSO Conference, and achieving progress in greater technical synergy between safety and security, progress must still be made in drafting the proposed safety guidance on TSOs. He also suggested that the IAEA consider prioritizing the drafting of such guidance. Because nuclear safety is science based, TSOs have a special role to play in addressing the safety challenges facing the nuclear community. TSO functions are part of each step of a holistic approach to nuclear safety. Guidance from international organizations such as the IAEA is needed in this regard. Recognition of TSO functions has progressed at the IAEA and worldwide (e.g. a European directive); progress is still needed. Scientific and technical capabilities are progressing but still need to be reinforced, and the support provided by TSOs is essential for embarking and emerging countries. The science based work in TSO functions requires research to develop the science and knowledge base. In addition, there is a need for worldwide harmonization of nuclear safety practices to meet the highest standards. Regional and international networks enhance the work of TSOs, and peer reviews — e.g. as part of Integrated Regulatory Review Service (IRRS) missions — can play an important role in this regard. Assistance to newcomers is a key part of harmonization needs. Mr. Repussard ended his keynote address by highlighting the vital role of communication, stating that the public and other stakeholders must be provided with science based information, as societal vigilance is essential to nuclear safety.

Gustavo Caruso, Special Coordinator for the IAEA Action Plan on Nuclear Safety, stated that approximately 70% of the nearly 800 tasks in the IAEA Action Plan on Nuclear Safety have been completed. Among the tasks where TSOs have provided key support are the safety assessments (stress tests); IAEA peer reviews; the extension of capabilities in RANET; the review of IAEA safety standards; and harmonization of the liability regime within the international legal framework. He further emphasized that open and transparent communication is crucial; to that point, the IAEA is taking a leading role in putting together the forthcoming report on the Fukushima Daiichi accident. The report will be a factual evaluation and assessment of the accident comprising a summary report, written to be understandable to laypersons, and five scientific/technical chapters. Finally, he emphasized that not all the lessons learned from the Fukushima Daiichi accident were new lessons and stressed that, when moving forward, steps need to be taken to ensure that past lessons 'stay learned'.

Masashi Hirano, Japan Director General for Regulatory Standard and Research, Nuclear Regulatory Authority (NRA) provided a summary of activities in Japan related to the Fukushima Daiichi accident, including details of the development of new regulatory requirements by the NRA, effective as of July 2013. In addition, he stated that a total of 20 nuclear power plants have applied for a conformance review for restart. He recalled that the Japanese Diet report had concluded that the lack of expertise was one of the fundamental causes of the Fukushima accident, further stating that 'technical independence' is an essential element of an effective nuclear regulator, along with political independence and financial independence. In this context, on 1 March 2014, the Japan Nuclear Energy Safety Organization (JNES) was merged into the NRA to enhance the latter's technical competence and expertise. At the same time, a new department was created in the NRA to serve as an 'internal TSO'. In parallel, cooperation with the Nuclear Safety Research Center of the Japan Atomic Energy Agency (JAEA) and the National Institute of Radiological Sciences (NIRS), which are the 'external TSOs', has been strengthened.

Regarding Fukushima Daiichi, the Tokyo Electric Power Company (TEPCO) has been conducting various activities according to the Mid-and-Long-Term Roadmap towards Decommissioning under the supervision of the Government's Council for Decommissioning. TEPCO has already completed more than 75% of the fuel removal from the spent fuel pool at Unit 4. The large amount of radioactive water being generated daily is a difficult issue that will need to be addressed through long term efforts. The highly radioactive water remaining in the seawater pipe trenches in the seaside area is currently the highest contributor to risk. TEPCO is attempting to plug the flow paths between the trenches and turbine buildings by applying an ice plugging technique, which will also be applied for the 'frozen soil wall' surrounding Units 1 to 4.

OVERVIEW OF TOPICAL ISSUE SESSIONS

Topical Issue 1: The Role of TSOs in relation to the Fukushima Daiichi Accident

The session presented challenges and solutions in TSO responses to the Fukushima Daiichi accident, TSO involvement in the implementation of stress tests, formulation and implementation of nuclear safety regulations, responses to the IAEA Action Plan on Nuclear Safety and work in post-accident recovery. Seven presentations were made by China, Japan, Germany, the Russian Federation, Slovenia, and the United Kingdom.

Topical Issue 2: Interface Issues

The session presented a wide array of challenges and issues that TSOs face when interacting with the regulatory body, industry and the public, as well as safety and security issues, during both non-emergency and emergency situations. Six presentations were made by France, the Republic of Korea, Finland, the Russian Federation, the IAEA and WANO.

Topical Issue 3: Emergency Preparedness and Response

The session discussed the roles of and challenges faced by TSOs in terms of emergency preparedness and response, assessment, prognosis and monitoring, as well as the regulatory and legislative frameworks in some countries that protect the TSO experts during an emergency response. Seven presentations were made by Canada, China, France, the Republic of Korea, Morocco, Ukraine and the IAEA.

Topical Issue 4: Maintaining and Strengthening TSO Capabilities

The session presented the challenges TSOs face in maintaining professional expertise, building capacity, and understanding the human and organizational factors that can affect both. Seven presentations were made by Belgium, Canada, China, France, the USA and the IAEA.

Topical Issue 5: Networking among TSOs and beyond

The session presented the issues and challenges that TSOs face, as well as the benefits they derive, when networking in a global environment. Six presentations were made by the Czech Republic, Lithuania, EU–JRC/EC (Joint Research Centre), the Ibero-American Forum of Radiological and Nuclear Regulatory Agencies (FORO), the IAEA and OECD Nuclear Energy Agency.

CLOSING SESSION

This session was held as a panel discussion and focused on future developments and on visions of future cooperation among TSOs. It agreed on the conclusions and recommendations of the conference.

CONCLUSIONS

1.   While much progress has been made — especially since the accident at the Fukushima Daiichi nuclear power plant in March 2011 — the lessons learned during this TSO Conference, much like some of the lessons learned from the Fukushima Daiichi accident, are not new. TSOs continue to face the same, ongoing challenges. However, the need to face these challenges has never been more important and will

continue to be so in the future. This is equally true for countries embarking on new nuclear power programmes, for countries with existing power reactors that are facing long term operation issues, and for countries that have decided to phase out nuclear power and are facing the associated tasks.

Like the first International Conference on the Challenges Faced by Technical and Scientific Support Organizations in Enhancing Nuclear Safety, held in 2007 in Aix-en-Provence, France, and the second, held in 2010 in Tokyo, Japan, this conference recognized that providing TSO experts from different countries and organizations with an opportunity to meet to discuss and develop a common understanding of their responsibilities, needs, risks and opportunities is one of the few mechanisms available for moving forward. TSO leaders and other stakeholders concluded that a fourth conference dedicated to these issues was needed, to be held in three or four years.

Recommendation 1: The IAEA should consider initiating plans for a fourth international conference on TSO functions, science and expertise. In this respect, the conference welcomed the proposal of Belgium to host the next TSO Conference.

2.   Both the development and the implementation of nuclear regulations are, and must be, science based. Decisions made by regulators with regard to nuclear safety and security are complex and require a thorough understanding of the science and technology underlying the radiation and nuclear activities being regulated. Regulators must have at their disposal a body of independent technical and scientific safety and security experts and advisors to support their primary development and the implementation of regulations.

There was general agreement that the regulatory body must: (1) maintain its authoritative function, in relation to nuclear facilities and other licensed activities involving radioactive substances, by having permanent access to a competent and sustainable technical and scientific advisory expert function; (2) provide for the continuing education, capacity building and knowledge sharing needs of these experts so that they may continue to provide competent advisory services to the regulator in a timely manner; and (3) recognize that the true costs associated with poor safety extend far beyond the money saved in cutting funding to R&D programmes, and that such cuts in reality undermine the nuclear safety and security framework itself.

It was further noted that, for countries with newly developed nuclear workforces and those with ageing workforces, there is a need for capacity building, mentoring, training and education to ensure nuclear safety and security, and to maintain a strong safety culture.

3.   There was strong consensus on the need to manage the interfaces between the TSO and stakeholders, the TSO and the regulator, and the TSO and industry — whether the TSO experts are situated within the organizations they provide services to or are external to them.

It was recognized that TSOs have a responsibility to contribute to addressing public and societal expectations with respect to communication to and involvement of the public, including providing the public with access to expertise. This is important for increasing public confidence, especially as the public generally has more confidence in experts.

The conference stressed that TSOs should give more attention to conducting research aimed at ensuring the safety and security of existing and future facilities and activities. Common research projects should be developed among different kinds of

TSOs, using existing frameworks and research networks to the extent possible. There were also discussions on understanding and managing the interfaces between safety and security at all levels.

4.   TSO functions are a critical component of regulatory systems.

The previous conference recognized that regulatory systems comprise three major functions: the regulatory body authoritative function, the technical and scientific expertise function and the function of developing the knowledge base and associated tools. The last two constitute the 'TSO functions'.

5.   The IAEA as a driving force: There is a critical need for guidance on TSO functions to support and enhance regulatory control.

There are large differences between countries with respect to technical approaches to performing TSO functions; it is therefore necessary to strengthen the safety and security framework through the scientific and technical expertise that supports it. Guidance from the IAEA is critical in this regard.

TSO functions require the following elements: a strong science and knowledge base, an integrated approach to assessment, and technical expert judgment. These elements are referred to in several IAEA Safety Requirements; ideally, however, they would be addressed in a Specific Safety Guide.

As noted at this and the 2010 TSO Conference, the safety requirements that address TSO functions are not presented as a cross-cutting reference in a single, comprehensive Safety Guide, nor are they covered in an effective manner across the body of Safety Guides. The conference reiterated the benefits of developing such a guide, both to those countries that already have nuclear programmes as well as to newcomers.

Recommendation 2: There was general agreement that the IAEA should consider producing a Safety Guide on the performance of TSO functions as part of the IAEA Safety Standards Series.

6.   Peer reviews are recognized as a key means of enhancing safety and security. Better coverage of TSO functions in peer review missions, such as IRRS missions, would allow Member States to benchmark and improve their capabilities.

The IRRS peer review service covers regulatory and organizational aspects of regulatory systems extensively, but does not address technical aspects in a thorough manner. This is due in part to the absence of adequate standards in this field in the IAEA safety standards. If such references were available in a Safety Guide, it would allow TSO functions to be adequately covered in IRRS missions, in order to allow Member States to benchmark and improve their capabilities.

Recommendation 3: The IAEA should consider including in IRRS or other peer review missions the evaluation of the capabilities of those national TSO functions that contribute to research, training and education, and the performance of safety assessments. It should also consider establishing specific missions or modules within existing missions to that effect.

7.   TSO needs and functions are particularly crucial for newcomers.

There was strong consensus on the need for newcomer countries, countries with expanding nuclear energy programmes and those phasing out their nuclear energy programmes to create, develop or maintain their scientific, safety and security capacities. As new regulatory authorities are being established, their need for support is growing. TSO functions are instrumental in helping to ensure the establishment of the necessary scientific and technical safety and security knowledge and capacity, which is an integral part of the regulatory processes. Yet, progress in this area has been limited since the 2010 TSO Conference.

Recommendation 4: There was general agreement that the IAEA should consider fully integrating the work of the TSO Forum to better support newcomer countries, as a valuable means of contributing to building their capabilities.

8.   Special efforts are needed to further develop and maintain the knowledge and competence base.

Effective nuclear safety and security supervision requires access to state of the art assessment capabilities. Such capabilities require scientific and technical knowledge in the nuclear field, together with sufficient assessment experience and proper management of knowledge and know-how. They also require maintaining advanced technical infrastructure, such as experimental facilities and computer codes, which can be shared among TSOs in order to pool resources and avoid duplication. They must be continuously developed, in a sustainable manner, at the national, regional or international level through:

- Scientific, risk oriented research;

- Relevant operating experience analysis, which is key to building the knowledge and expertise database;

- Knowledge management, dissemination and transfer to new generations of experts;

- Professional training courses and tutoring.

In particular, R&D is indispensable in creating and developing the safety and security knowledge and expertise required for the assessment of existing and future facilities. There still exist knowledge gaps that require research, especially in view of the continuous development of nuclear technology, and this undermines the credibility of regulatory assessments.

Recommendation 5: Common nuclear safety research projects should be developed among organizations carrying out TSO functions (1) using existing frameworks to the extent possible, in particular those provided by the IAEA and the OECD/NEA, and (2) through other efficient means, such as by creating or joining regional TSO networks.

9.   The existing knowledge and capabilities of Member States are not sufficiently reflected in the combined IAEA Secretariat and Member State emergency response capabilities.

In this respect, the conference concluded that TSOs should register their capabilities in the IAEA Response and Assistance Network (RANET).

Recommendation 6: TSOs should make every effort to coordinate with the authorities in their countries in order to register their capabilities within the IAEA Response and Assistance Network (RANET).

Recommendation 7: The IAEA should consider further integrating TSO expertise, to enhance the capabilities for assessment and prognosis during nuclear or radiological emergencies.

10.   There is a widely shared concern that developing and maintaining TSO capacities currently is not always adequately resourced — or that it may not be in the future.

Ensuring long term planning and funding is critical to the effectiveness and sustainability of TSO functions. TSO functions must have the appropriate expertise and adequate resources, and there is concern that this situation is not ensured to the same degree in all countries.

The conference pointed out that organizations performing TSO functions must have the resources to maintain independence of judgement while also achieving the highest level of technical competence and transparency. They should thus be able to provide independent technical and scientific advice without pressure from regulatory bodies, industry or other stakeholders.

Recommendation 8: Member States should ensure that adequate and sustainable resources are available to maintain the TSO capacity.

11.   The IAEA Action Plan on Nuclear Safety has created momentum among Member States, and the TSO Forum plays an important role as a driving force for continued support.

The conference highlighted several positive examples of results achieved by the TSO Forum. There was general agreement that international cooperation and networking among TSOs is crucial, as it contributes to increasing experience feedback and provides the experience and the information base needed to tackle new cases.

The conference recognized that the IAEA, through the TSO Forum, is a strong driving force for the development of TSO functions and capabilities.

Recommendation 9: The IAEA should consider further expanding the activities of the TSO Forum, to develop it into a science and expertise forum providing comprehensive coverage of issues concerning TSO functions in nuclear regulation. It should also consider establishing, in particular, new means of improving international networking to share knowledge and experience on technical and scientific practices.

ACKNOWLEDGEMENTS

The IAEA wishes to thank all the hosts and participants who contributed to the conference. Upon the successful completion of the conference, the IAEA would like to thank the Government of China for hosting the conference in Beijing, for their outstanding organization and provision of excellent facilities, and for their hospitality extended to everyone involved in the conference (participants, IAEA staff, speakers and chairpersons). The IAEA would also like to thank the TSOs from China, in particular the Nuclear and Radiation Safety Centre (NSC) of the National Nuclear Safety Administration (NNSA) for their commitment in making this conference such a success. Special thanks and appreciation go to Ms Feng Yi from the NSC as the point of contact for the IAEA for her involvement and valuable support prior to and during the conference. IAEA would also like to extend further thanks to Ms Cui Dandan from the Ministry of Environmental Protection of China as the focal point who committed herself to ensure the smooth running and successful outcome of this conference. Their professionalism and kind support were very much appreciated. The IAEA also would like to thank the European Technical Safety Organisation Network (ETSON) for its excellent cooperation.

CONFERENCE ORGANIZATION

PRESIDENT OF THE CONFERENCE

B. De Boeck   Belgium

Co-President

Z. Li   China

CHAIRPERSONS OF SESSIONS

Moderator of the Opening Session	P. Woodhouse	IAEA
Session 1 – The Role of TSOs in relation to the Fukushima Daiichi accident	M. Hirano	Japan
Session 2 – Interface issues	S. West	United States of America
Session 3 – Emergency Preparedness and Response	L. Bolshov	Russian Federation
Session 4 – Maintaining and strengthening TSO capabilities	A. dela Rosa	Philippines
Session 5 – Networking among TSOs and beyond	M. H. Kim	Republic of Korea

SECRETARIAT OF THE CONFERENCE

Scientific Secretaries: P. Woodhouse, IAEA
                                  L. Guo, IAEA
                                  M. Heitsch, IAEA
                                  (until August 2014)

Local Organizers: D. Cui, China

Administrative Support: Z. Zohori IAEA
                                  T. Maier IAEA

PROGRAMME COMMITTEE

Chair: B. De Boeck	Belgium
T. Jamieson	Canada
D. Cui	China
J. Cheng	China
M. Heitsch	EU/EC-JRC
E.K. Puska	Finland
J.-B. Chérié	France
R. Dallendre	France
C. Eibl-Schwaeger	Germany
L. Guo	IAEA
P. Woodhouse	IAEA
K. Tomita	Japan

CONTRIBUTORS TO DRAFTING AND REVIEW

Guo, L.	IAEA
Ben Ouaghrem, K.	IAEA

80

Annex

CONTENTS OF THE ATTACHED CD-ROM

The attached CD-ROM contains the technical programme of the conference, the list of participants, all available papers as well as presentations and posters provided during the conference. The "CN-214 Book of Synopsis and structure of sessions" document reflects all contributions in the order they were given at the conference. The "CN-214 Book of all contributions" lists the abstracts of all contributions in the order they were registered.

IAEA
International Atomic Energy Agency

ORDERING LOCALLY

In the following countries, IAEA priced publications may be purchased from the sources listed below or from major local booksellers.

Orders for unpriced publications should be made directly to the IAEA. The contact details are given at the end of this list.

CANADA

Renouf Publishing Co. Ltd

22-1010 Polytek Street, Ottawa, ON K1J 9J1, CANADA
Telephone: +1 613 745 2665 • Fax: +1 643 745 7660
Email: order@renoufbooks.com • Web site: www.renoufbooks.com

Bernan / Rowman & Littlefield

15200 NBN Way, Blue Ridge Summit, PA 17214, USA
Tel: +1 800 462 6420 • Fax: +1 800 338 4550
Email: orders@rowman.com Web site: www.rowman.com/bernan

CZECH REPUBLIC

Suweco CZ, s.r.o.

Sestupná 153/11, 162 00 Prague 6, CZECH REPUBLIC
Telephone: +420 242 459 205 • Fax: +420 284 821 646
Email: nakup@suweco.cz • Web site: www.suweco.cz

FRANCE

Form-Edit

5 rue Janssen, PO Box 25, 75921 Paris CEDEX, FRANCE
Telephone: +33 1 42 01 49 49 • Fax: +33 1 42 01 90 90
Email: formedit@formedit.fr • Web site: www.form-edit.com

GERMANY

Goethe Buchhandlung Teubig GmbH

Schweitzer Fachinformationen
Willstätterstrasse 15, 40549 Düsseldorf, GERMANY
Telephone: +49 (0) 211 49 874 015 • Fax: +49 (0) 211 49 874 28
Email: kundenbetreuung.goethe@schweitzer-online.de • Web site: www.goethebuch.de

INDIA

Allied Publishers

1st Floor, Dubash House, 15, J.N. Heredi Marg, Ballard Estate, Mumbai 400001, INDIA
Telephone: +91 22 4212 6930/31/69 • Fax: +91 22 2261 7928
Email: alliedpl@vsnl.com • Web site: www.alliedpublishers.com

Bookwell

3/79 Nirankari, Delhi 110009, INDIA
Telephone: +91 11 2760 1283/4536
Email: bkwell@nde.vsnl.net.in • Web site: www.bookwellindia.com

ITALY
Libreria Scientifica "AEIOU"
Via Vincenzo Maria Coronelli 6, 20146 Milan, ITALY
Telephone: +39 02 48 95 45 52 • Fax: +39 02 48 95 45 48
Email: info@libreriaaeiou.eu • Web site: www.libreriaaeiou.eu

JAPAN
Maruzen-Yushodo Co., Ltd
10-10 Yotsuyasakamachi, Shinjuku-ku, Tokyo 160-0002, JAPAN
Telephone: +81 3 4335 9312 • Fax: +81 3 4335 9364
Email: bookimport@maruzen.co.jp • Web site: www.maruzen.co.jp

RUSSIAN FEDERATION
Scientific and Engineering Centre for Nuclear and Radiation Safety
107140, Moscow, Malaya Krasnoselskaya st. 2/8, bld. 5, RUSSIAN FEDERATION
Telephone: +7 499 264 00 03 • Fax: +7 499 264 28 59
Email: secnrs@secnrs.ru • Web site: www.secnrs.ru

UNITED STATES OF AMERICA
Bernan / Rowman & Littlefield
15200 NBN Way, Blue Ridge Summit, PA 17214, USA
Tel: +1 800 462 6420 • Fax: +1 800 338 4550
Email: orders@rowman.com • Web site: www.rowman.com/bernan

Renouf Publishing Co. Ltd
812 Proctor Avenue, Ogdensburg, NY 13669-2205, USA
Telephone: +1 888 551 7470 • Fax: +1 888 551 7471
Email: orders@renoufbooks.com • Web site: www.renoufbooks.com

Orders for both priced and unpriced publications may be addressed directly to:
Marketing and Sales Unit
International Atomic Energy Agency
Vienna International Centre, PO Box 100, 1400 Vienna, Austria
Telephone: +43 1 2600 22529 or 22530 • Fax: +43 1 2600 29302 or +43 1 26007 22529
Email: sales.publications@iaea.org • Web site: www.iaea.org/books

18-04041